METHODS IN
MEMBRANE BIOLOGY

VOLUME 1

Contributors to This Volume

A. D. Bangham, *Biophysics Unit, A. R. C., Institute of Animal Physiology, Babraham, Cambridge, England*

S. Ferrone, *Department of Experimental Pathology, Scripps Clinic and Research Foundation, La Jolla, California*

N. L. Gershfeld, *National Institute of Arthritis, Metabolism, and Digestive Diseases, National Institutes of Health, Bethesda, Maryland*

M. W. Hill, *Biophysics Unit, A. R. C., Institute of Animal Physiology, Babraham, Cambridge, England*

Yasuo Kagawa, *Department of Biochemistry, Jichi Medical School, Kawachi-gun, Japan*

M. M. Long, *Laboratory of Molecular Biophysics, University of Alabama Medical Center, Birmingham, Alabama*

N. G. A. Miller, *Biophysics Unit, A. R. C., Institute of Animal Physiology, Babraham, Cambridge, England*

M. A. Pellegrino, *Department of Experimental Pathology, Scripps Clinic and Research Foundation, La Jolla, California*

R. A. Reisfeld, *Department of Experimental Pathology, Scripps Clinic and Research Foundation, La Jolla, California*

D. W. Urry, *Laboratory of Molecular Biophysics, University of Alabama Medical Center, Birmingham, Alabama*

METHODS IN MEMBRANE BIOLOGY

VOLUME 1

Edited by EDWARD D. KORN

Section on Cellular Biochemistry
and Ultrastructure
National Heart and Lung Institute
Bethesda, Maryland

PLENUM PRESS • NEW YORK–LONDON

Library of Congress Cataloging in Publication Data

Korn, Edward D.   1928-
   Methods in membrane biology.

   Includes bibliographies.
   1. Membranes (Biology) I. Title. [DNLM: 1. Membranes — Periodicals.
W1 ME9616C]
QH601.K67   574.8'75   73-81094
ISBN 0-306-36801-3

© 1974 Plenum Press, New York
A Division of Plenum Publishing Corporation
227 West 17th Street, New York, N.Y. 10011

United Kingdom edition published by Plenum Press, London
A Division of Plenum Publishing Company, Ltd.
Davis House (4th Floor), 8 Scrubs Lane, Harlesden, London, NW10 6SE, England

Articles Planned for Future Volumes

Nuclear Magnetic Relaxation and the Biological Membrane
  A. G. Lee, N. J. M. Birdsall, and *J. C. Metcalfe*
  (National Institute for Medical Research, London)

Kinetic Studies of Transport Across Red Blood Cell Membranes
  Y. Eilam and *W. D. Stein* (The Hebrew University of Jerusalem)

Formation of Impermeable Inside-Out and Right-Side-Out Vesicles from
  Erythrocyte Membranes
  Theodore L. Steck (University of Chicago)

Isolation and Characterization of Membrane Glycosphingolipids
  Roger A. Laine, Klaus Stellner, and *Sen-itiroh Hakomori*
  (University of Washington)

Isolation and Characterization of Surface Membrane Glycoproteins from
  Mammalian Cells
  Mary Catherine Glick (University of Pennsylvania)

The Use of Monolayer Techniques in Reconstitution of Biological Activities
  Lawrence Rothfield (University of Connecticut)

Techniques in the Formation and Examination of "Black" Lipid Bilayer
  Membranes
  D. A. Haydon, R. Fettiplace, L. G. M. Gordon, S. B. Hladky, J. Requena,
  and *H. P. Zingsheim* (Cambridge University)

Isolation of Plasma Membranes from Mammalian Cells
  David M. Neville, Jr. (National Institutes of Health)

The Isolation and Characterization of Gap Junctions
  Daniel A. Goodenough (Harvard Medical School)

The Determination of Protein Turnover in Biological Membranes
  Robert T. Schimke, David Shapiro, and *John Taylor* (Stanford University)

Cell Fractionation Techniques
  H. Beaufay and *A. Amar-Costesec* (Universite Catholique de Louvain)

X-Ray Diffraction of Biological Membranes
  J. Kent Blasie (University of Pennsylvania)

Lectins in Membrane Biology
  H. Lis and Nathan Sharon (Weizmann Institute of Science)

Fluorescence Measurement in Membrane Biology
  G. K. Radda (Oxford University)

Preparation and Characterization of Isolated Intestinal Epithelial Cells and
  Their Use in Studying Intestinal Transport
  George Kimmich (The University of Rochester)

Electrochemical and Optical Methods for Studying the Excitability of the Nerve
  Membrane
  I. Tasaki, Emilio Carbone, and *Ken Sisco* (National Institutes of Health)

Methods for Studying Transport in Bacteria
  Thomas H. Wilson, Eva Kashket, and *Peter Maloney* (Harvard Medical
  School)

Methods for Determining the Topographical Distribution of Proteins in
  Membranes
  Martin Morrison (St. Jude Children's Research Hospital)

The Use of Isolated Membrane Vesicles in Transport Studies
  Joy Hochstadt (The Worcester Foundation for Experimental Biology)

Methods for Molecular Weight Estimation of Membrane Proteins and
  Polypeptides
  Wayne W. Fish (Medical University of South Carolina)

Methods of Isolation and Characterization of Bacterial Membranes
  Milton R. J. Salton (New York University Medical Center)

Polypeptide Hormone-Receptor Interactions: Quantitative Aspects
  C. R. Kahn (National Institute of Arthritis, Metabolism, and Digestive Diseases)

Preface

Examination of the tables of contents of journals — biochemical, molecular biological, ultrastructural, and physiological — provides convincing evidence that membrane biology will be in the 1970s what biochemical genetics was in the 1960s. And for good reason. If genetics is the mechanism for maintaining and transmitting the essentials of life, membranes are in many ways the essence of life. The minimal requirement for independent existence is the individualism provided by the separation of "life" from the environment. The cell exists by virtue of its surface membrane. One might define the first living organism as that stage of evolution where macromolecular catalysts or self-reproducing polymers were first segregated from the surrounding milieu by a membrane. Whether that early membrane resembled present cell membranes is irrelevant. What matters is that a membrane would have provided a mechanism for maintaining a local concentration of molecules, facilitating chemical evolution and allowing it to evolve into biochemical evolution. That or yet more primitive membranes, such as a hydrocarbon monolayer at an air–water interface, could also have provided a surface that would facilitate the aggregation and specific orientation of molecules and catalyze their interactions.

If primitive membranes were much more than mere passive barriers to free diffusion, how much more is this true of the membranes of contemporary forms of life. A major revolution in biological thought has been the recognition that the cell, and especially the eukaryotic cell, is a bewildering maze of membranes and membranous organelles. Parallel with this development has been the realization that these membranes are not just the static barriers and demarcators of cellular space they may appear to be in stark photomicrographs. Rather, membranes are dynamic structures in continual movement, both morphological and molecular. Moreover, the major biochemical and physiological events of life occur in, on, or through membranes. Not just the highly selective, often energy-coupled, transport of ions and molecules but also the complex processes of oxidative phosphorylation, photosynthesis, vision, and nerve conductance; intermediary biochemical events such as protein and lipid biosynthesis, the citric acid cycle, and fatty acid oxidation; hormone–cell and cell–cell interactions; endocytosis and secretory processes are all membrane phenomena. It is at the membrane that morphology and metabolism unite; that catalytic chaos is organized; that self is distinguished from nonself.

Elucidation of the structure, function, and biosynthesis of membranes thus becomes a major goal of contemporary experimental biology. The tasks are to determine the chemical and enzymatic constituents of the membranes; to study the physical and chemical properties of relevant models; to dissociate the biological membranes into their natural structural and functional units; to recombine these minimal units into membranes that are structurally and functionally identical to the originals; to discover the biological mechanisms of synthesis and organization; to understand the varied roles of membranes in normal and pathological states. In practice, of course, all of these approaches are undertaken simultaneously, and progress in any one area is a tremendous stimulus to success in the others.

Unfortunately, the methodological as well as the conceptual difficulties are immense. Morphological and ultrastructural techniques are inherently static and in any case stop short of revealing information at the molecular level. Physical and chemical methods are averaging techniques providing considerable information about the mean properties of membranes, but only a general guide to the organization of the specific and diverse functionally distinct subregions within them. Traditional enzymology has many weapons with which to fight in the sea of aqueous reactions but possesses few tools with which to dig into the fertile fields of surface chemistry. Despite the enormous difficulties, however, considerable progress is being made in all areas by the use of a wide variety of old and new methods.

The generality of the importance of membranes and the tremendous diversity of experimental approaches to their study create yet another difficulty which it is the purpose of this series to help overcome. In membrane research perhaps more than in any other area of biology, progress depends on methodology, and the methodology can be highly technical and highly specialized. As a consequence, one investigator often finds it difficult to understand and to evaluate the techniques used by another. Authors for *Methods in Membrane Biology* have been urged, and given adequate space, not only to describe their methods in sufficient detail for the reader to use them in his laboratory (or at least to tell the reader where he can find the experimental details when they are readily available elsewhere) but also to discuss fully the theoretical backgrounds of the methods and their applications and limitations in membrane biology. The expressed aim is to enable each of us to evaluate more critically, and to understand more fully, data obtained by methods foreign to our usual experiences. It is planned that each volume will contain a range of methods varying from the physical to the physiological, thus maximizing the audience for each. "Methods" may at times be interpreted rather broadly, but it is not intended that these articles shall be primarily reviews of the results obtained by the methods under discussion.

It is entirely appropriate that Volume 1 of *Methods in Membrane Biology* should begin with a chapter on liposomes. It is now generally accepted that, whatever the degree of complexity of the arrangement of proteins and carbohydrates in membranes, most of the phospholipids of biological membranes are in molecular bilayers (although there still is debate on the extent of uninterrupted lipid bilayer within some membranes), for which the liposomes are simple and elegant experimental models. Indeed, the liposome may resemble even more closely that putative primitive membrane referred to earlier. In any case, whether or not it is an evolutionary as well as an experimental prototype, the liposome is a fascinating system for structural and functional studies of phospholipid bilayers. Although myelin figures had been known for many years and lipid dispersions had been studied previously, it was A. D. Bangham who first made explicit use of the liposome as a model membrane. In their chapter, Bangham, Hill, and Miller present a succinct and authoritative review of the preparation of multilamellar and single-bilayer liposomes and of their structural and functional applications in membrane biology.

Ultimately, our understanding of lipid interactions will be complete only when it rests on a thermodynamically sound support. N. Gershfeld has modified the experimental approach to the Langmuir trough and performed simple, definitive experiments with lipid monolayers. The data are analyzed so as to reach conclusions that are always interesting and frequently provocative. The surface chemist contributed very much very early to membrane theory, and he still has much to offer the biologist.

At a more complex level of organization, the third chapter deals with spectroscopic analysis of membrane proteins, where it is particularly important to recognize the experimental difficulties. D. Urry was among the first to realize that, valuable though optical rotatory dispersion and circular dichroism measurements were in providing information about the configurations of membrane proteins, the particulate nature of the systems imposed severe limitations on the interpretation of the spectra. Urry and Long have now developed theoretical and experimental ways to circumvent many of these difficulties, allowing a fuller utilization of spectroscopy in studies of membrane structure.

The next chapter in this volume discusses the physiologically, pathologically, and therapeutically vital subject of the antigenic properties of the cell surface. Here, major experimental problems are to design quantitative assays for membrane antigens and to develop methods for the isolation and purification of surface antigens from normal and abnormal cells. Reisfeld, Ferrone, and Pellegrino describe their and others' efforts with respect to the histocompatibility antigens in this rapidly developing field.

Finally, the chapter by Y. Kagawa is a comprehensive presentation of the brilliant achievements in the structural and functional reconstitution of that most complex and integrated system, the inner mitochondrial membrane. While dealing specifically with the dissection and reassembly of the energy transfer reactions of the mitochondrion, Kagawa has developed a set of general principles and techniques that should prove of inestimable value as a rational guide for investigations of all biological membranes.

It is our intention that future volumes in this series will appear at not greater than annual intervals. At the time of this writing Volumes 2, 3, and 4 are at various stages of gestation; their anticipated contents are listed separately. Future volumes will best serve their intended purposes if readers will take the time to communicate to the editor "methods" that they would like to have critically reviewed. If the reader can also suggest the names of possible authors, that would be of additional value. Should they wish to volunteer themselves, so much the better, but acceptance is not guaranteed for then the editor would have to do nothing but read their proofs and collect his royalties.

September, 1973

Edward D. Korn

Contents

Chapter 2

Thermodynamics and Experimental Methods for Equilibrium Studies with Lipid Monolayers

N. L. GERSHFELD

Chapter 3

Circular Dichroism and Absorption Studies on Biomembranes

D. W. URRY AND M. M. LONG

Chapter 4

Isolation and Serological Evaluation of HL-A Antigens Solubilized from Cultured Human Lymphoid Cells

R. A. REISFELD, S. FERRONE, AND M. A. PELLEGRINO

Chapter 5

Dissociation and Reassembly of the Inner Mitochondrial Membrane

YASUO KAGAWA

Chapter 1

Preparation and Use of Liposomes as Models of Biological Membranes

A. D. Bangham, M. W. Hill, and N. G. A. Miller

Biophysics Unit, A.R.C.
Institute of Animal Physiology
Babraham, Cambridge, England

1. INTRODUCTION AND HISTORICAL SURVEY

The recognition that biological cells exploit the water–oil interfacial activity of certain lipids to define anatomical boundaries has, in recent years, encouraged many workers in this laboratory* and others to develop and study protein-free model membrane systems prepared with such compounds. A considerable technical advance was made more than 10 years ago, when Mueller *et al.* (1962a,b) first reported a method for preparing usable preparations of bimolecular membrane from membrane molecules, the so-called black lipid membranes or BLMs. The merits of this powerful technique were quickly realized, and, as with Aladdin's lamp, a property requested was a property acquired! Indeed, it might be said that the skeptics themselves were the holders of the lamp, because it was they who, by dismissing the model as unrealistic for a succession of reasons, indicated explicitly the characteristic implicit (passive) properties of a biological membrane: before very long, the earlier convictions of the Langmuir–Hardy–Rideal schools of surface

---

* We borrow Kinsky's connotation of the word "laboratory" as a means of indicating chronologically the names of colleagues who have participated in the development of the liposome model: J. C. Watkins, M. M. Standish, G. Weissmann, D. Papahadjopoulos, J. de Gier, G. D. Greville, S. L. Bonting, S. M. Johnson, R. C. MacDonald, R. Klein, M. Moore, M. A. Singer, P. Callissano, R. Lester.

chemistry, relating the role of amphiphilic molecules to biological membrane structure, were vindicated.

Unlike the mechanically supported black lipid membranes, contrived for the first time so very recently, lyotropic smectic mesophases (liposomes) of membrane-like lipids have probably been forming and reforming on the aqueous earth for longer than life itself. Indeed, it is worth remarking, yet again, on the similarity which might exist between the present state of model membrane syntheses and some stage in prebiotic history (Bangham, 1968, 1972). Such a statement might not seem a very flattering assessment of present-day achievements, but in reality it takes us a very considerable distance along the evolutionary trail. Simple principles of electrostatic, hydration, and surface free energies now recognized as being the principal forces governing membrane stability could just as well have applied to a primitive population of compounds, of increasing carbon chain length, acquiring polar or ionic head groups and being synthesized in a continuum of water (Oparin, 1924).

The late A. S. C. Lawrence (1969), in a lifetime spent on this subject, described many ordered-phase systems of simple mixtures of long-chain alcohols, long-chain acids, and water. Together with Dervichian (1964), he recognized the physical similarities of phospholipids to their own simpler analogues. Leathes, too, as was pointed out by Small (1967), demonstrated that lecithins interact physically with water to form what were termed "myelin figures," which were recognized then as being "made up of films two lecithin molecules thick, with the hydrophilic groups facing the water on each surface" (quoted by Small, 1967).

An evolutionary sequence would have involved a selective adsorption of amphiphilic compounds at the air–water interface, their orientation to form a monolayer, and their eventual collapse due to aero- or hydrodynamic stresses to form closed aqueous compartments, i.e., liposomes. A (molecular) form of natural selection, based on surface free energy minima, would ensure the evolution of increasingly more stable bilayers. In the event of a particularly favorable synthetic process of membrane molecules evolving *within* an extant membrane system, a clone of membranes would develop. A factor reinforcing such speculation is the surprisingly long time constant for a molecule on one side of a bilayer to flip to the other (Kornberg and McConnell, 1971). The sequestered, aqueous compartments so formed would become isolated from the bulk primeval soup by a continuous, thin hydrocarbon membrane which would exhibit its very distinctive permeability properties, to be referred to later. The interfacial region, too, between aqueous phase and thin hydrocarbon film, would present a mosaic of hydrophilic groups such as hydroxyl, carbonyl oxygen, and amide as well as positive or negative ions. Most probably, such a planar array of ordered reactive sites would have both catalytic and template properties. There is also the very real and interesting possibility that simple polypeptides of the type which are now recognized as being selective ion carriers, coexisting in the primeval soup, might have initiated the familiar pattern of an intracellular enrichment of K^+ over Na^+.

In December 1932, a British subject (J. Y. Johnson, 1932) applied for a British patent on behalf of I.G. Farbenindustrie Aktiengesellschaft, who had found that "pharmaceutical preparations for injection into the muscular system or subcutaneously can be prepared by combining medicaments with liquids, such as fats or fatty oils, if necessary together with waxes or wax-like substances, with water, or other liquids, and a dispersing agent, whereby a system, hereinafter called "depot" is found capable of holding any desired doses of the medicament but releasing it over any desired space of time only gradually ... without the slightest detriment to the organism"! One specification particularly singles out lecithins as being a useful lipoid, and the description of a "depot" to contain strophantin reads singularly like a contemporary preparation of liposomes: "An emulsion is prepared from: 25 parts of lecithin; 20 parts of water; 1.5 parts of cholesterin; 0.03 parts of strophantin and 0.5 parts of a 'Nipasol' (p-hydroxybenzoic acid normal propyl ester)"; its success can now be ascribed to the fact that at least part of the dose of the drug had been sequestered within a smectic mesophase or liposome system. Unaware of the existence of this patent specification, one of the present authors, while preparing aqueous suspensions of lecithin for electron microscopy (Bangham, 1963; Bangham and Horne, 1964), observed by light microscopy that the structures could be seen to alter shape in response to changes of concentration of solute—whether of electrolytes or nonelectrolytes—in the continuous aqueous phase. Furthermore, it was noted that the presence or absence of charged phospholipids in the system modified the birefringence of the smectic mesophase at a given electrolyte concentration. These observations were followed up and subsequently reported as a possible complementary model to the BLM (Bangham *et al.*, 1965a,b, 1967a,b). Rendi, too, had noted that "suspensions" of total mitochondrial lipids in an aqueous phase "extruded" water in an analogous manner to mitochondria when exposed to calcium or albumin solutions (Rendi, 1965), but he seemed unaware of the membranous nature of the lipid material.

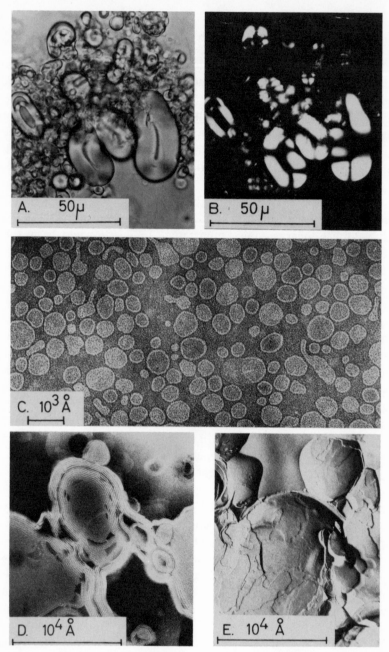

Fig. 1. **Microscopic appearances of liposomes.** A: Light microscopy, low power. Multilayered liposomes of phosphatidylcholine in 100 mM NaCl. B: Same field but with crossed polaroids. C: Microvesicles negatively stained in ammonium molybdate. D: Multilamellar liposomes in ammonium molybdate; from Bangham and Horne (1964). E: Multilamellar liposomes by freeze-fracture; from Deamer *et al.* (1970).

Finally, in this section it would be appropriate to define the term "liposome."* For the purpose of this chapter, it is understood to refer to all assemblages of phospho- and other lipids sustaining a bimolecular configuration but which by themselves do not require mechanical support for their stability. There are three principal types, namely, multilamellar, microvesicular, and macrovesicular (Fig. 1). As the succeeding sections will reveal, each type of liposome has been developed more for reasons of experimental expediency than from a curiosity as to whether the properties of the model itself are altered by size and/or shape. It would not surprise the present authors to learn, for example, that microvesicles are either more or less permeable than multilamellar systems per unit area of membrane; indeed, there have already been persistent reports that the shape of microvesicles cannot be altered by withdrawing water osmotically.

2. MODEL SYSTEMS

2.1. Multilamellar Liposomes

It was with preparations of multilamellar liposomes, observed under the microscope, that the inherent potentialities of smectic mesophases as model membranes were first recognized. The original observation was simply that a mesophase appeared to change shape when the concentration of bulk-phase solute was changed. For example, a few micromoles of mixed brain or egg lipids dissolved in chloroform can be placed, as a smeared-out drop, on the bottom of shallow 100-ml beaker. The chloroform is removed by evaporation with N_2 or air, and the beaker is placed on a microscope stage. While dry, the smear of lipids will appear as a colorless mass unless crossed polarizers and a first-order compensator are incorporated into the optical path; a ×10, or better a ×20, objective with a long working distance is recommended, when the anisotropic liquid crystalline lecithin will show spectral colors due to the random, domain structure and birefringence of the material. If now the smear is gently covered with a thin layer of a dilute salt solution, say, 10 mM, smectic mesophases will rapidly form, and the observer will be beguiled by the process. However, if after a few minutes one adds a quantity of a saturated salt solution to the beaker and mixes, rapid

* The multilayered versions were charmingly first referred to as "bangosomes" by Lehninger (1967).

shape changes will be seen to take place. Finally, one can remove by suction most of the hypertonic liquid from the beaker while keeping the same field in view and replace it once more with a large volume of the original (10 mM) swelling fluid.

It became crucial to establish, first, that the shape changes were a consequence of water moving *across* a semipermeable membrane and, second, that the forces were osmotic and not electrostatic. Identical behavior with dilute and concentrated nonelectrolytes, e.g., glucose, eliminated the latter consideration. Although the evidence presented by Bangham *et al.* (1965a,b) and Papahadjopoulos and Watkins (1967) was more compatible with a closed membrane than a Swiss roll or Schwann cell structure, it was not until valinomycin had been shown to facilitate selectively the diffusion of K^+ over Na^+ from liposomes containing equal concentrations of K^+ and Na^+ that the closed membrane theory could be finally claimed as proven (Bangham *et al.*, 1967a).

One of the attractions of using the multilamellar liposome model is that they are exceedingly easy to prepare; indeed, it is no exaggeration to say that they form themselves spontaneously! All that is required is that a sufficient quantity of a stock solution in chloroform of the phospholipid and/or sterols be placed in a round-bottomed flask and the solvent chloroform removed *in vacuo* in a rotary evaporator so that the lipids are left as a thin film on the surface of the flask. The aqueous phase, containing any labeled solute, is then added, together with a glass bead or two, and the flask is shaken. One recognizes that lyotropic mesomorphisms have taken place, i.e., multilamellar liposomes have formed, by noticing that the lipid film whitens and disperses into the excess aqueous phase. Useful quantities of ingredients for diverse experimental purposes may be obtained by reference to papers by Bangham *et al.* (1965a,b), (1967a,b); de Gier *et al.* (1968); S. M. Johnson and Bangham (1969a,b); Klein *et al.* (1971); Sessa and Weissmann (1970); Weissman and Sessa (1967). An idea of the diversity of the compounds that have been used to form liposomes may also be gained by a perusal of papers by Papahadjopoulos and Bangham (1966), who made liposomes with phosphatidylserine; by Papahadjopoulos and Miller (1967), who prepared liposomes with a variety of acidic phospholipids; by Demel *et al.* (1968), who used liposomes of natural and synthetic phosphatidylcholines with and without cholesterol; and by Lester *et al.* (1972) and Hill and Lester (1972), who examined the effects of introducing gangliosides into phosphatidylcholine liposomes. Failure to produce characteristic liposomes, assuming that appropriate lipid mixtures have been used, is most likely because the predominant lipid is below its transition temperature (see Section 3.4) and is not at all penetrated by water and solute. The emphasis on distributing the lipid out of chloroform as a *thin film* is related to a less than desirable phenomenon for sizable amounts of dry phospholipid become wrapped up within a concentric system of solute-permeated lamellae, the closed membranes effectively preventing further ingress of both water and salt.

Multilamellar liposomes, as models for study, have been criticized for a number of reasons (S. M. Johnson and Bangham, 1969*a,b*), sometimes, but not always, with justification. Undeniably, the most valid criticism is that it is not easy to prepare a suspension of liposomes having a uniform size population; thus the kinetics of efflux are more complex than the uniform "solid-sphere" (Carslaw and Jaeger, 1947) model first discussed by Bangham *et al.* (1965*a*) because of the magnitude of the size range. The recent work of Chowhan *et al.* (1972), however, seems to indicate that the heterogeneity of these preparations is not of great importance.

A doubtful criticism, based on an interpretation of freeze-fractured preparations, is that the multilamellar, coplanar structures contain inclusion spaces of considerable volume. It is true that the amount of a given solute sequestered is greater in multilamellar liposomes than in microvesicles per micromole of membrane components, but this is because the outer surface area of the sonicated microvesicles is as much as five or six times that of the hand-shaken preparation. As the outermost surface area increases, so will the proportion of membrane molecules facing outward. Indeed, the small discrepancy in trapping ratios between multilamellar and microvesicular liposomes, when related to the proportion of lipid molecules in the outermost skin, may be accounted for by the consensus that microvesicles are usually spherical with a minimum surface/volume ratio, whereas the multilamellar liposomes are oblate or prolate cylinders and/or spheroids.

Despite these criticisms, the multilamellar liposome can be used as a helpful and informative model system for the simple reason that not all experiments require absolute permeability coefficients. Thus, by normalizing a response, e.g., release of a labeled permeant or rate of swelling, to a control aliquot of a given preparation of liposomes, a great deal of information can be gained. From the literature, it is evident that the model has been useful in a variety of studies: for example, effect of surface charge on cation permeability (Bangham *et al.*, 1965*a*) and susceptibility to hydrolysis by phospholipases (Bangham and Dawson, 1958, 1959, 1962); the properties of ion carriers (Henderson *et al.*, 1969); the effect of lytic reagents, including some polyene antibiotics (Weissmann and Sessa, 1967); entrapment of enzymes into liposomes (Sessa and Weissmann, 1970; Gregoriadis *et al.*, 1971); the binding of soluble basic proteins such as cytochrome *c* (Kimelberg and Papahadjopoulos, 1971*a,b*) and soluble basic proteins such as mellitin (Sessa *et al.*, 1969); interaction with crystals of monosodium urate (Weissmann, 1971; Weissmann and Rita, 1972), with albumin (Sweet and Zull, 1969), with steroids (Sessa and Weissmann, 1968), with antibiotics (Kinsky, 1970), with retinal and retinaldehyde (Bonting and Bangham, 1967), with thyroid hormones (Hillier, 1970), with lyso- and short-chain phosphatidylcholines (Reman *et al.*, 1969), and with chlorophyll (Chapman and Fast, 1968; Trosper and Raveed, 1970); interaction with basic polypeptides (Hammes and Schullery, 1970); complement lysis of erythrocytes (Kinsky, 1972); action of local and general anesthetics (Papahadjopoulos, 1972); effect of membrane composition on small nonelectrolyte permeabilities (de Gier *et al.*, 1968, 1969, 1971); differential aminoacid permeabilities (Klein *et al.*, 1971); differential nonelectrolyte permeabilities (Cohen and Bangham, 1972); and total salt permeabilities (Singer and Bangham, 1971).

Furthermore, a technique has been developed for measuring the surface area of the outer (only) bimolecular membrane (see Section 4), enabling the calculation of true permeabilities (Bangham *et al.*, 1967*b*).

2.2 Microvesicles

Interest in the microvesicle structural version of the smectic mesophase has arisen for a number of reasons. For example, Saunders and his group (Saunders *et al.*, 1962; Gammack *et al.*, 1964) were motivated by a pharmaceutical requirement to "solubilize" cholesterol. They found (Saunders *et al.*, 1962) that ultrasonic irradiation of phospholipids in an atmosphere of nitrogen would give rise to stable, optically clear sols, provided the hydrocarbon chains were of an appropriate length or degree of unsaturation. The implications of this early work, together with that of Lawrence (1969) pertaining to the temperature of water penetration into fatty acid and fatty alcohol mixtures, are only now being appreciated. Our laboratory, too, became interested in Saunders' preparations because, when mixed with dilute aqueous solutions of sodium phosphotungstate, they yielded excellent electron microscopic images (Bangham, 1963; Saunders, 1963; Bangham and Horne, 1964). Indeed, it was from these early negatively stained images that the idea of the entity of closed membranes developed. Looking back over the prints, taken some 10 years ago, it is quite clear that in some of the

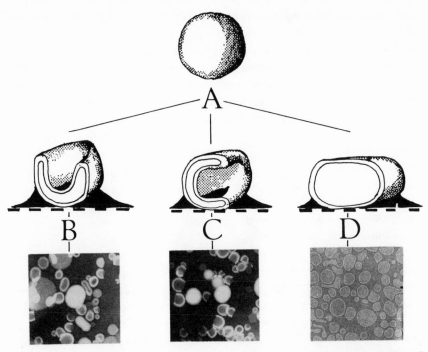

Fig. 2. Schematic representation of events which might occur during negative staining of spherical, single-bilayer sonicated liposomes to give rise to the electron micrographs. The top row illustrates a spherical liposome in suspension. The second row illustrates configurations the liposomes might assume on the electron microscopic grid in the absence of albumin (B and C) and in the presence of albumin (D). In each case, the liposome has been bisected to reveal the configuration of the single-bilayer membrane. The bottom row illustrates the electron microscopic image. From S. M. Johnson *et al.* (1971).

preparations, but not in others, the negative stains penetrated into the vesicle. An artifact which was to deceive us for some time was the apparent "double"-membraned microvesicle (see Fig. 2, bottom, taken from S. M. Johnson *et al.*, 1971). Figure 2 was ultimately offered by the authors as a possible explanation.

Abramson *et al.* ultrasonicated brain phosphatidylserine (1964*a*) and phosphatidic acid (1964*b*) in aqueous media because they wanted to study the ionic structure and ionic exchange capabilities of purified acidic lipids. However, their data were interpreted at the time as though the phosphatidylserine molecules were aggregated as spherical pincushion micelles. Papahadjopoulos and Watkins (1967), in a wider ranging study of the smectic mesophase of membrane phospholipids, came to the conclusion that phosphatidylserine and phosphatidylcholine formed microvesicles when sonicated. They pointed out that a vesicular structure need not contradict the results obtained by Abramson et al., because rupture of the vesicles would be expected to occur following the changes of pH.

S. M. Johnson and Bangham (1969a) switched to using ultrasonicated smectic mesophases of egg phosphatidylcholine and egg phosphatidic acid in an effort to obtain true permeability coefficients and to simplify kinetic analysis (see later section). Apart from misinterpreting the microscopic evidence and inadvertently introducing a compensating error, the exercise was worthwhile, because the preparations and their measurements achieved both objectives to a certain extent. However, the present authors are alert to the possibility that the constraints on the membrane of the microvesicle are more severe than on multilamellae, possibly with some consequent alteration of properties.

Huang (1969) introduced the method of molecular sieve chromatography on Sepharose 4B columns for the purpose of obtaining a completely uniform population of microvesicles. Huang's procedure, in effect, separated residual multilamellar liposomes as a void volume fraction of the bead bed, recovering ultimately a population of vesicles small enough to occupy the internal volume of the gel as fraction II (Fig. 3). Although these preparations are, by all criteria, rather homogeneous, it requires to be shown that the prolonged and intimate exposure of the vesicles to a solid surface does not alter their properties; one wonders whether a brief ultracentrifugation (Seufert, 1970) might not achieve the same result more rapidly. Finer et al. (1972) have followed the kinetics of sonication in some detail. For example, Fig. 4, taken from their paper, illustrates the time course of sonication as measured by either high-resolution nuclear magnetic resonance spectroscopy or absorbance at 300 nm. Analysis of these data showed that the process was second order.

Microvesicles may be produced in two ways, namely, solvent evaporation or ultrasonication of multilamellar smectic mesophases. Solvent evaporation (Robinson, 1960; Papahadjopoulos and Watkins, 1967), curiously enough, has not proved a popular method, and the present authors are not sure why. It consists, as the name implies, of layering a solution of the phospholipids in an organic solvent, e.g., petroleum ether (60–80°C portion), onto the aqueous phase and then removing the petroleum ether with a stream of O_2-free N_2. Presumably, as yet, no practical criteria exist to establish the degree to which the solvent molecules have been removed from the system; any residue would be quite unacceptable. Furthermore, there

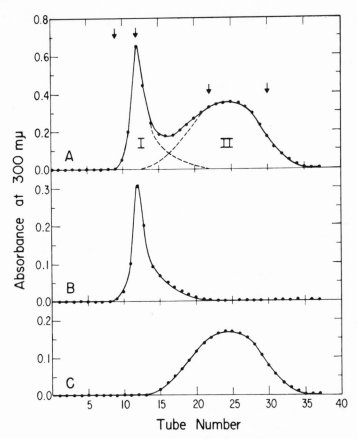

Fig. 3. Elution patterns of phosphatidylcholine. A: Dispersion obtained from ultrasonic irradiation. Phosphatidylcholine dispersion (8 ml of 3%) was applied to a 2.5- by 50-cm lipid-treated Sepharose 4B column. B: Vesicles collected and concentrated from fraction I of part A as indicated by the first two arrows. C: Vesicles collected and concentrated from fraction II of part A as indicated by the last two arrows. From Huang (1969).

would be uncontrolled loss of water from the aqueous phase with a con-sequent increase in solute concentration. A more recent method has been described by Chowhan *et al.* (1972) in which an appropriate volume of a 15% chloroform solution of phospholipids is stirred directly with the permeant-labeled, aqueous phase under a gentle reduced pressure; subsequently, the chloroform is removed under vacuum. The authors claim that this method yields a uniform population of multilamellar liposomes giving predictable efflux profiles.

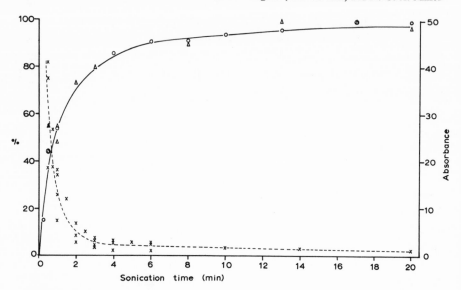

Fig. 4. Time course of sonication. Percent lecithin molecules in fraction II, Fig. 3, (O———O), percent N(CH₃)₃ groups giving high-resolution NMR integral ($\triangle$———$\triangle$), and absorbance at 300 nm ($\times$———$\times$) as a function of sonication time of a 1 % (w/v) dispersion. The solid curve through the experimental points for percent molecules in fraction II and giving a high-resolution spectrum is calculated according to the theory of Finer *et al.* (1972) described in the text.

Ultrasonication presents technical difficulties, too, not least that of being able to measure, and therefore meter, the amount of energy actually being imparted to the smectic mesophases. Since it has been a consistent, and in our view correct, policy to avoid immersion of a metallic probe into the aqueous phase, we have, of necessity been concerned with effective energy transfer. Metallic probes certainly deliver the energy where it is required, but they erode and the fragments of the metal contaminate the microvesicular preparation; more seriously, they are likely to dissolve and to promote oxidative degradation. Hauser (1971), using such a probe, claims that prolonged exposure at low intensity or short exposure at high intensity causes appreciable chemical degradation, yielding lysolecithin, fatty acids, glyceryl phosphorylcholine, and phosphorylcholine. Because it is also desirable to sonicate in an atmosphere of nitrogen and to provide a heat sink, we advocate the use of a simple waterbath sonicator (Kerry's KB 80/1). The sample, say, up to 50 μmoles in 1.0 ml, is sealed in a glass or plastic ampule after thorough flushing with nitrogen and located in the focal point of radiant energy within the bath. The resonant volume depends on the geometry of the bath and quantity of water in it. The bath temperature can be controlled by draining unwanted heat away via water circulating in a copper coil. The most important trick of all, however, is to reduce the surface tension of the water in the bath to minimize the dissipation of sonic energy by cavitation of the water in the bath itself; any good aqueous detergent is effective.

Oxidation during ultrasonication is always a hazard, particularly when the vulnerable, polyunsaturated lipids are being used. Saha *et al.* (1970) noted that microvesicles, for example, prepared from cardiolipin, were particularly sensitive to degradation by oxidation because of the high percentage of unsaturated fatty acids. Huang (1969) monitored the development of absorption maxima between 265 and 285 nm in the ultraviolet spectrum, but, as Klein (1970*a*) points out, the appearance of such maxima indicates peroxidation of fatty acids containing only three or more double bonds (i.e., linolenate and higher), whereas the increase in diene conjugation as judged by absorbance changes at 233 nm would measure changes in linoleate as well as the higher polyunsaturated chains. Klein reported a simple technique for detecting oxidation in microvesicular preparations based on the absorption ratio: A_{233}/A_{215}. The technique merely requires the addition of 3 ml absolute ethanol to 0.1 ml aqueous phase containing approximately 2 μmoles of phospholipid and reading absorbance at 233 nm and 215 nm. The lower limit of detection was conservatively estimated as 0.02 absorbance unit, equivalent to 0.1% oxidation, assuming reasonable values for the relevant extinction coefficients.

It is common experience that for phospholipids, pure or as mixtures, above their transition temperature, some are more resistant to ultrasonic energy than others. In practice, lecithin alone or with cholesterol is found to be the most difficult material to homogenize, entirely consistent with its molecular dimensions and physical properties. The presence of a functionally wedge-shaped molecule—molecularly steric or because of an ionizable group—always seems to facilitate homogenization. Phospholipids, pure or as mixtures, below their transition temperature do not form smectic mesophases (Dervichian, 1964).

Microvesicles have been exploited to measure, among other parameters, the ionic structure and ion exchange capabilities of biological phospholipids (Abramson *et al.*, 1964*a,b*); differential permeability to various ions (Papahadjopoulos and Watkins, 1967; Papahadjopoulos, 1971); absolute permeabilities to K^+ at various temperatures with and without valinomycin and anesthetics (S. M. Johnson and Bangham, 1969*a,b*);

interactions with basic proteins (Kimelberg and Papahadjopoulos, 1971a,b); dielectric studies (Redwood et al., 1972); activation of adenosine triphosphatase (Wheeler and Whittam, 1970); the reversal effect of pressure on anesthesia (S. M. Johnson and Miller, 1970); inside–outside transition of phospholipid molecules (Kornberg and McConnell, 1971); the thickness of a lipid bilayer in an excess aqueous phase (Wilkins et al., 1971); the partial specific volume, diffusion, and sedimentation coefficients and molecular weight of phospholipids (Saunders et al., 1962; Huang, 1969; S. M. Johnson and Buttress, 1972); the refractive index of microvesicle bilayers (Seufert, 1970); high-resolution NMR spectroscopy (Lee et al., 1972; Penkett et al., 1968; Kaufman et al., 1970; Birdsall et al., 1971; Metcalfe et al., 1971); the partition coefficients and effects of certain steroids (Heap et al., 1970, 1971).

2.3. Macrovesicles

In 1969, Reeves and Dowben reported an interesting variation on the smectic mesophase model. We here suggest that they be called "macrovesicles" despite the fact that they are purported to be about 1 μ in diameter! They are formed as follows: a chloroform–methanol (1:2) solution of egg lecithin is taken to dryness on a very flat-bottomed flask; the amount of phospholipid (5 μmoles) is small and requires to be thinly distributed on the bottom of a 2-liter flask. Thereafter, the dry lamellae of phospholipid are exposed to water-saturated nitrogen until they, too, become swollen and saturated with water. Parenthetically, this stage of the preparation would correspond to that X-rayed by Levine and Wilkins (1971), where a multilayer periodicity of 51.5 Å was measured, equivalent to a "wet" region of 17.6 Å. Finally, a 0.2 M sucrose solution at 42 °C is gently added to the flask, which is then left standing for a few hours.

The important property of the vesicles, once formed and harvested, is that their walls consist of only a few bimolecular sheets. Thus they rather nearly resemble a biological cell or cell organelle. However, for reasons which are not clear, the vesicles do not form when electrolyte or solute protein is added in the place of the nonelectrolyte. It should, however, be possible to treat the vesicles as though they were erythrocytes and to lyse and reseal them with selected marker solutes. Reeves and Dowben (1970), using their macrovesicle, report some elegant stop-flow measurements on osmotic water permeability that are consistent with the view that water permeates by dissolution and diffusion in the bimolecular membrane(s).

3. PHYSICAL PROPERTIES OF MEMBRANE MOLECULES IN AQUEOUS MEDIA

Membrane molecules are characteristically amphiphilic, which is to say that they are sufficiently large for different regions of each molecule to behave in a discrete fashion: thus each molecule behaves as though it were oil soluble at one end and water soluble at the other. Typical examples of membrane compounds which exhibit amphiphilic properties to the extent that they form highly ordered phases in equilibrium with a water phase (liposomes) are the class II polar lipids (Small, 1970), e.g., phosphatidylcholines, -ethanolamines, and -serines, sphingomyelins, cardiolipins, plasmalogens, phosphatidic acids, and cerebrosides. These compounds, when confronted with an encroaching aqueous environment, may be observed to undergo a sequence of assemblages which reflect the thermodynamic perturbations of increasing water–water, water–oil, and oil–oil interactions. Kavanau (1965) analyzed these thermodynamic considerations and accounted for the paradoxical emergence and sustained existence of highly ordered structures from within a solution. These quasi-equilibrium structures, colloquially referred to in the title of this chapter as "liposomes," should, more correctly, be termed "smectic mesophases." Smectic mesophases, whether formed gently on a microscope slide or harshly by swirling lipids and an aqueous phase together, are layer lattices of alternating bimolecular lipid sheets, intercalated by aqueous spaces. When formed from water-insoluble amphiphiles such as those mentioned above, they persist even in the presence of excess water. Their usefulness as a model system derives from the fact that, as the dry phospho- and/or other lipids of biological origin undergo their sequence of molecular rearrangements, there is an opportunity for an unrestricted entry of solutes, e.g., isotopically labeled salts and proteins, between the planes of hydrophilic head groups before the unfavorable entropy situation of an oil–water interface intervenes. Subsequently, and because of the unfavorable entropy associated with the excess aqueous phase, a further arrangement of molecules takes place which yields a series of concentric closed membranes, each membrane representing an unbroken, bimolecular sheet of molecules. From this, it follows that every aqueous compartment is discrete and isolated from its neighbor by one such closed membrane and that the outermost aqueous compartment of the whole structure (Fig. 1C) would be isolated from the continuous aqueous phase. The solutes and water originally entering the system are therefore sequestered and can only diffuse between compartments or into the bulk aqueous phase by crossing one or more bimolecular membranes.

3.1. Microscopy

A truly fascinating pastime is to place a drop of water on a dried deposit consisting of mixed membrane lipids and then observe the sequence of events under the microscope. Greater insight as to the degree of ordering of the structures formed is instantly obtained if a polarizer and analyzer are placed at right angles above and below the specimen. For these weakly birefringent structures (Fig. 1B), it also helps to introduce a first-order red compensator or near equivalent in the form of a sheet (0.75 mm) of mica. Although crude mixtures of membrane lipids may be the most readily available material for the above pleasantries, they do not exhibit as strong a birefringence as preparations made with pure samples of, say, dioleoyl phosphatidylcholine or egg lecithin.

The sign and magnitude of the birefringence exhibited by smectic mesophases, as with other crystals, depend on the sum of two components: an intrinsic component which may be positive or negative and a form component (due to the parallel array of lamellae intercalated by sheets of water) which is negative. The intrinsic birefringence is a characteristic of the species and orientation of the individual molecules, whereas the form component can and does vary in magnitude as a function of the relative thickness and refractive index of the two parallel lipid and aqueous phases. If the equilibrium distance between two adjacent lamellae depends on the balance of attractive and repulsive forces, absence of any net charge, as, for example, with pure lecithin, will result in close apposition and strong positive birefringence. The presence of a net charge, whether positive or negative, will cause a separation of adjacent lamellae and an intensification of the negative-form birefringence, ultimately resulting in an overall loss or even reversal in the sign of the observed birefringence (Bangham et al., 1965a). Such a relationship between the degree of swelling (hydration) of lipids in electrolytes arises because of the interplay of fixed charges and ionic strength, and was first pointed out by Palmer and Schmitt (1941) using X-ray diffraction studies of mixed brain lipids. More recently, one of the present authors together with colleagues (Bangham et al., 1965a,b, 1967a) has confirmed in a qualitative fashion the correlation between the optical appearances, the sequestered and pellet volumes, the X-ray spacings, and the fixed charge densities of smectic mesophases Figure 5, for example, illustrates the difference between the

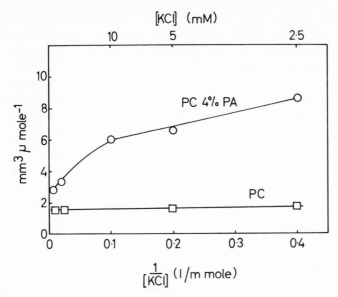

Fig. 5. Difference in pellet volumes after mixing with KCl to form liposomes. Innate pellet volumes (mm³/μmole of lipid) of dispersions of phosphatidylcholine (PC) (□) and of phosphatidylcholine–phosphatidic acid (PA) (○) allowed to form in various concentrations of KCl. From Bangham *et al.* (1967*b*).

pellet volume of a micromole of pure phosphatidylcholine and a micromole of a mixture of phospholipids containing a charged component (such as might be present in a mixture of membrane lipids) after mixing with a range of concentrations of KCl to form liposomes. The pure phosphatidylcholine, with no net charge, had an invariant pellet volume, exhibited a constant intensity of birefringence, and gave a constant lamellar spacing.

Thus not only are individual membrane molecules shuffled around by favorable energetic forces and/or unfavorable entropic changes, but longer-range forces ensure that the evolving lamellae of the smectic mesophase are accurately spaced. Precisely for these reasons, electron microscopy of phospholipids in aqueous media poses rather special problems. Bangham and Horne (1964), in discussing this dilemma, suggested that negative stains might offer several important advantages for the direct visualization of water-sensitive, lipid phases. First, the rigid setting of the salt (e.g., sodium phosphotungstate, ammonium molybdate) in the form of an electron-dense "glass" is believed to preserve structures as though they were in the presence of water; second, lipid solvents are not used; and, finally, a very high degree of resolution can be obtained. On the other hand, Bangham and Horne (1964) were at the time all too aware of some of the artifacts associated with the technique (Glauert and Lucy (1969). For example, it was noted that adsorbing multivalent ions, e.g., phosphotungstate, caused distortion of the structure by electrostatic stress—a technical point learned during earlier efforts to negatively stain erythrocyte ghosts, which always disintegrated in sodium or potassium phosphotungstate but which were instantly recognizable in calcium phosphotungstate or ammonium molybdate. The preferential adsorption of phosphotungstate but not molybdate may be measured by microelectrophoretic techniques (see Section 2). Notwithstanding these reservations, the method has stood the test of time, for, in expert hands and correctly interpreted, valid images have been obtained (Bangham, 1963; Glauert and Lucy, 1969; Papahadjopoulos and Miller, 1967; Junger and Reinauer, 1969; Junger et al., 1970; Tinker and Pinteric, 1971; S. M. Johnson et al., 1971) (Fig. 1C,D). But it would always be prudent to examine material in as many different negative stains as possible (sodium phosphotungstate, calcium phosphotungstate, and ammonium molybdate) as well as by freeze-fracture and freeze etching. These latter techniques, likewise, are audacious in assuming that the prevailing molecular structure is "frozen" before the consequences of dehydration are attained. Elegant freeze-fractured images of liposomes have been published by Fluck et al. (1969) and by Deamer et al. (1970) (Fig. 1E). It would certainly be the opinion of the present authors that the freeze-fracture technique will ultimately offer the most revealing and candid images of these delicate structures!

3.2. X-Ray Diffraction

A trouble with X-ray diffraction techniques is that they give but average information: a time exposure of many hours is necessary when in reality the subject(s) are fidgeting about in nanoseconds. It is probable that the first X-ray diffraction of liposomes was made in St. Louis by Schmitt et al. in 1935, for in their own words "the dried solid obtained by evaporation of the benzene extract of cow spinal cord was rubbed up with water and placed on the stirrup (sample holder) in the moist chamber before the pinhole. The patterns contained rings at 4.16 and 15.7 Å, both spacings observed in fresh medulated nerve ... Similar results were obtained also with lecithins and with a cholesterol–lecithin mixture." Whereas single smectic mesophases give relatively sharp Bragg reflections for both short and long spacings (Finean and Millington, 1955; Luzzati and Husson, 1962; Small, 1967; Levine and

Wilkins, 1971), smectic mesophases in the presence of a second water phase give diffuse reflections because there is a tendency for the stacked arrays of membranes to be randomly orientated; there is a further complication because of the highly variable long spacing between each fragment of mesophase (Bangham *et al.*, 1967*b*). Such limitations have always applied to biological objects, but fortunately there have been exceptional structures such as spinal cords and retinal rod outer segments. Recently, Wilkins *et al.* (1971) have reported on a diffraction analysis of dispersions of membranes, including liposomes, not in regular arrays. They argue that for a random dispersion of sheets, the diffraction pattern represents a continuous-intensity distribution of rings centered on the direction of the incident beam. Diffraction, in effect, is spread out more as θ (Bragg angle) increases, and the square of the Fourier transform (F^2) of the sheet profile becomes proportional to $I \sin^2$, where I is the intensity for angle θ. When sheets form small spherical shells, e.g., sonicated phospholipids, interference between X-rays diffracted by different parts of the shell produce a modulation of I, but apparently the modulation is unlikely to obtrude unless the vesicles are extremely uniform in size. Figure 6 illustrates the curve of $I^{0.5} \sin \theta$ *vs.* $\sin \theta$ for a 15% lecithin-in-water sonicated dispersion. The main diffracting band corresponds to a sheet of thickness $D \approx 36$ Å, i.e., the interglycerol distance of a bimolecular layer.

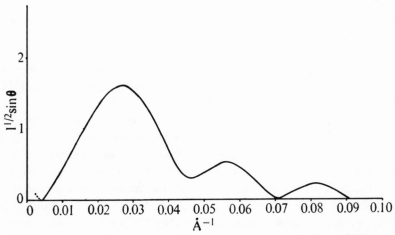

Fig. 6. Diffraction from sonicated dispersion of egg lecithin in water. From Wilkins *et al.* (1971).

3.3. Light Scattering

The use of light-scattering techniques to investigate the physical properties of smectic mesophases in all states has two attractive attributes: namely, it is nondestructive and the changes can be followed rapidly. However, great complexities arise in the interpretation of the results, as it is difficult to disentangle the effects of asymmetry, size, and heterogeneity. For a detailed study on all aspects of light scattering, the reader is referred to a comprehensive review by Kerker (1969). The three types of liposome dispersions have to be considered separately in relation to their isotropic light-scattering properties. First are the "clear solutions" of single-walled microvesicles with diameter of the order of $\lambda/20$, second the large single-walled macrovesicles with diameters in the range 1–20 μ, and finally multilamellar preparations which exhibit a very large size range.

Microvesicles were first studied systematically by Saunders *et al.* (1962), who, interested in their intrinsic stability, showed that a suspension of egg lecithin decreased in optical density as it was progressively irradiated with ultrasonic energy. They found no further decrement of optical density beyond about 20 min sonication, and the dispersion thus formed was stable for up to 77 hr. The clearest solution had an optical density of the order of 0.1 for a 5 % dispersion by weight. No attempt was made to use these data to obtain a size determination. Abramson *et al.* (1964*a*) investigated the ultrasonicated dispersion of the sodium salt of phosphatidylserine and obtained a molecular weight of 6×10^6 by the use of the Zimm plot and a dissymmetry ratio of 2.27–2.38. Their molecular weight is in reasonable agreement with determinations made by centrifugation, but the asymmetry factor is excessively high, giving a diameter for a sphere approximately of 2000 Å, or of more than 4000 Å for the length of a rod-shaped micelle. This discrepancy appears again in the work of Attwood and Saunders (1965), which is referred to later.

The scattering of light by these small vesicles is best interpreted in terms of the Rayleigh–Debye theory. The theory gives the scattering as a product of two terms: the scattering, assuming the particles are Rayleigh scatterers, times a form factor which is dependent on the size and shape of the particles. It is the form factor that gives the information from which the dimensions of the particles may be adduced. Form factors can be constructed from hypothetical models, and these are then compared with experimental data. Seufert (1970), for example, analyzed data obtained on sonicated "asolectin"* microvesicles in this way by assuming that the particles are spherical shells the inside and the outside of which are identical. Using certain approximations as to the polarizability of the lipid regions, a plot of the dissymmetry parameter for shells of various thicknesses was obtained and compared with the experimentally determined dissymmetry, which led him to conclude that the diameter of the sphere was 400 Å. By carefully varying the refractive index of the aqueous medium, a measure of the effective refractive index of the lipid was obtained and given as 1.46. This agrees with that obtained by Cherry and Chapman (1969) on black lipid films. Using a similar preparation, but this time the vesicles were selected by column chromatography, Miyamoto and Stoeckenius (1971) showed that the suspension had no specific absorption between 650 and 300 nm and that the turbidity varied as the fourth power of the wavelength. Zimm plots were also employed, and a radius of gyration and molecular weight were found that agreed fairly well with the values given by Seufert (1970). The dissymmetry parameter is very sensitive to particle size (and shape) and, as Seufert points out, could be a useful technique in measuring volume changes of the particles under osmotic shock, if indeed they can change their shape.

There is broad agreement between the work referred to above and the data obtained by other techniques, but this could be more apparent than real, for the work of Attwood and Saunders (1965) on lecithin dispersions and the recent analysis by Tinker (1972) show that light-scattering techniques have to be treated with care. For example, Attwood and Saunders (1965) obtained molecular weights from Zimm plots of 2×10^6, in agreement with other work, but a reciprocal scattering factor more in agreement with that expected from thin rods than with that of the spheres they are shown to be by other techniques. The very full theoretical treatment presented by Tinker (1972) only goes part of the way to explain this discrepancy. One point of interest in comparing the Zimm plots of Attwood and Saunders and of Miyamoto and Stoeckenius is that the former have horizontal zero-angle lines, indicating little interaction of the particle with the solvent, whereas the latter has a finite slope. This could be due to the charges on the "asolectin," which as has already been pointed out, is a mixture of many lipids. It should

---

* To quote from Seufert (1970): "'Asolectin' is a commercially available preparation of soybean phospholipids manufactured by Associated Concentrates Inc. (Woodwide, Long Island, New York). It contains mainly phosphatidyl cholines, phosphatidyl ethanolamines and monophosphoinositides, small amounts of lysolecithin, phosphatidic acid and diphosphatidyl glycerol are also present. The main fatty acids in the preparation are linoleic acid (58%), oleic acid (10%) and palmitic acid (24%). The phosphorus content on a dry weight basis is 3%."

be noted that if there is any heterogeneity in a sample, dimensions obtained from light-scattering and other methods will be appropriate averages of that quantity. For instance, if the radius of gyration $\langle R_g \rangle$ is measured from a Zimm plot, this is in fact averaged as

$$\langle R_g{}^2 \rangle = \frac{\sum_i f_i M_i^2 (R_g)_i^2}{\sum_i f_i M_i^2}$$

where f_i is the frequency of the ith species and M_i its mass. If, on the other hand, the mean radius $\langle R \rangle$ is determined microscopically, this will be given by

$$\langle R \rangle = \frac{\sum_i f_i R_i}{\sum f_i}$$

These are not the same value even for a sphere, as M_i will be different for different values of $\langle R_g \rangle_i^2$.

The second case to be considered is that of the large single-layered macrovesicles. Reeves and Dowben (1970) used light-scattering techniques to measure the water permeability of these, their own type of vesicle. Complications arise due to a large heterogeneity, but a relatively simple theory is confirmed by the dependence of the optical density of a suspension being proportional to the second power of the wavelength. By assuming a simple form for the net refractive index of the whole vesicle as a function of its volume, the relative optical density before and after shrinking could be analyzed as a volume change, and hence, by following the time course of the optical density change, the permeability could be measured.

The final case, that of the multilayers, has not been theoretically studied, but similar empirical relationships to those obtained by Tedeschi and Harris (1955) for intact mitochondria have been shown to hold for these structures by Bangham et al. (1967b) and Rendi (1967). This relationship shows that the total volume of the scatterers is proportional to the reciprocal of the extinction coefficient. A direct measurement of this is shown in Fig. 7, where the total volume of the lipid plus sequestered aqueous phase is found after centrifugation and plotted as a function of the reciprocal of the extinction. An indirect result is shown in Fig. 8, where the multilayers were shrunk by hypertonic solutions of impermeable solutes and the reciprocal of the extinction coefficient is shown to be linearly related to the inverse of the concentration (or osmolarity). This fact has proved extremely useful, for it is possible to measure rapid changes of volume by this method in a

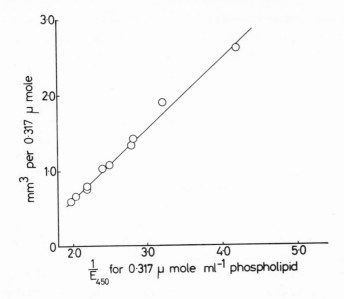

Fig. 7. **Pellet volumes (mm³/0.317 μmole of phospholipid) plotted against the reciprocals of the extinctions at 450 mμ for dispersions at a concentration of 0.317 μmole/ml.** Phospholipid composition as in Fig. 1. The points include innate volumes and volumes of swollen dispersions (SP800 spectrophotometer). From Bangham *et al.* (1967*b*).

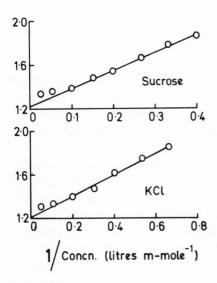

Fig. 8. Relationship, at equilibrium, between reciprocal of extinction at 450 mμ and reciprocal of solute concentration. A mixture of phosphatidylcholine and dicetyl-phosphoric acid (molecular proportions 99:1) was used, and the stock dispersion was formed in water. This was diluted with mannitol, sucrose, or KCl solutions of various concentrations. Phospholipid concentration in cuvettes, 0.5 mM (SP500 spectrophotometer). From Bangham *et al.* (1967*b*).

light-scattering apparatus or spectrophotometer and hence obtain permeability coefficients for water and other small soluble nonelectrolytes (Cohen and Bangham, 1972), details of which are given in later sections.

3.4. Microcalorimetry

Information on the thermal properties of a system lead immediately to the possibility of extracting thermodynamic data from that system. It is most important to realize that thermodynamics can only be applied to well-characterized "pure" systems and that information on the free energy is only a very meager part of that information. It is much more informative to have information on the derivatives of the free energy than the free energy itself, as these are more sensitive to subtle changes and should be used when comparing a real system with its model. After these words of warning, it is necessary to say that qualitative information of importance can nevertheless be gathered by calorimetric techniques, and with a system as well defined as the smectic mesophase, when pure known lipids are used, good thermodynamic data can be obtained.

Calorimeters fall into two categories, adiabatic and isothermal. The first is simpler to construct and use but thermodynamically difficult to interpret. The differential thermal analysis (DTA) consists of two sample holders, one containing the sample and other a blank. Both are heated at the same rate, but when the two samples have different heat capacities the one with the higher capacity will become cooler than the other, and in the event of an isothermal transition, i.e., melting, the temperature of the sample will remain constant while that of the control will rise. For measuring transition temperatures, the system is extremely accurate, but, as the sample temperature is varying all the time, the heat capacity or "latent heat" of a transition is difficult to determine. These instruments are commercially available, and Ladbrooke and Chapman (1969) refer to the use of one such model in detail.

The isothermal system called the "differential scanning calorimeter" (DSC) has been more extensively used. Here the two samples are kept at the *same* predetermined temperature (which is normally programmed to vary linearly with time) by varying the relative heats supplied to each sample. If the temperature is raised linearly and the heat difference plotted as a function of temperature, the height of the curve above baseline (with no sample each side, or systems kept at constant temperature) will be the difference between the specific heats of the sample and reference. At an isothermal transition, the area under the peak will be the "latent heat" of that transition. The temperature at which the transition occurs is difficult to locate accurately but conventionally can be taken as the first point of "liftoff" from the baseline.

The Perkin Elmer Corporation DSC 1B, now superseded by the DSC 2 (Watson *et al.*, 1964; O'Neill, 1964), has been used by many laboratories to study the phase transitions that occur in smectic mesophases. Ladbrooke and Chapman (1969) describe the use of the system in some detail, to which must be added a few practical details. As the properties of the lipids will depend on their concentration in the aqueous phase, it is necessary to use sealed pans. These are made of aluminum and have a volume of 20 μl. Ideally, there should be no air space above the sample, as this would result in water condensing out of the lipid solution onto the upper surface; however, in practice it is difficult to fill the pan with more than, say, 15 μl, as any lipid on the lip of the pan will result in a failure of the pan to seal. The pans should be cleaned before use in lipid solvents, and care should be taken not to damage the sealing edge. If the pan fails to seal, this will be obvious, as the heat necessary for evaporation will be recorded as a large drift. The amount of lipid necessary will depend on the experiment, but for transition temperature measurements the minimum quantity that is detectable is ideal, about 1 mg of dipalmitoyl lecithin, for example. For latent heats, the maximum concentration of lipid is nearly always necessary, say, about 5 mg of lecithin. The instrument is very sensitive to the purge gas flow rate. Anything that changes this will lead to erratic results, as water tends to accumulate in the gas flow system and the O-ring seals have a tendency to freeze and leak if used below ambient temperatures for long periods.

Danforth *et al.* (1967) have given details of a very sensitive form of the DSC which uses volumes of the order of 1.7 ml. The temperature difference between the two samples is held to ± 25 μ°C, and they claim to be able to measure heats of transition of 50 cal/liter of solution taking place over a 30°C range to 5% accuracy, and heat capacity changes of 2 cal/deg/liter of solution to 15% accuracy.

DSC curves for the gel-to-liquid crystalline transition of 1,2-distearoyl-DL-phosphatidylcholine with various water concentrations show that at high water concentrations there are two peaks, one due to the lipid and one to the water. At molar ratios of 10 water to 1 lipid, the water peak disappears and the transition temperature moves to higher values. Mixtures of two different phosphatidylcholines give one melting peak when made up together in organic solvent, but two when dispersed separately in water and then mixed. Most natural membranes and lipids consist of a large range of phospholipids of differing alkyl chain lengths and number of double bonds; this causes the transition peak to be spread out over a large range and also lowered so as to be undetectable. In the case of egg yolk phosphatidylcholine, there is a small peak at -15 to $-5\,^\circ$C, but it is impossible to say if this is just one species solidifying or if the system is supercooling.

Lipids extracted from *Mycoplasma laidlawii* grown on special media have proved to have definite phase transitions which are mirrored by those which can be measured with the natural membrane, an approach studied by various authors (Melchior *et al.*, 1970; Reinert and Steim, 1970). The absence of the transition in most natural membranes does not mean that the thermodynamic state of the hydrocarbon is irrelevant. It has been shown by various authors that the degree to which the hydrocarbon chains are unsaturated affects the permeability of natural membranes and that smectic mesophases made with comparable lipids exhibit similar deviations (de Gier *et al.*, 1968; McElhaney *et al.*, 1970; Klein *et al.*, 1971).

3.5. Centrifugation

More evidence of the size and shape of the sonicated vesicles has been gathered from the analytical ultracentrifuge. Saunders *et al.* (1962) investigated the sedimentation of "lecithin sols," then called, and obtained diffusion coefficients in the range 1.8×10^{-7} to 8×10^{-8} cm^2·sec^{-1} over a 26-hr period, indicating some heterogeneity of their preparation. Huang (1969) continued this earlier work, obtaining a diffusion coefficient D_{20}^0 of 1.87×10^{-7} cm^2 sec^{-1} and a sedimentation coefficient $S_{20,w}^0$ of 2.10. His fraction II (see Section 2.2) was a very homogeneous preparation of microvesicles giving a heterogeneity parameter of 0.5×10^{-14} sec compared with $\pm 5 \times 10^{-14}$ sec for unfractionated preparations. His conclusions were that the molecular weight of the vesicles was 2.06×10^6, with a Stokes' diameter of 228 ± 5 Å.

Working with microvesicles prepared with 4% phosphatidic acid, 96% phosphatidylcholine, S. M. Johnson *et al.* (1971) were able to reconcile a Stokes' radius obtained for the total vesicle from diffusion coefficients and the value for the radius of a sphere required to accommodate the measured amount of sequestered solute, e.g., ^{42}K. The measured external area of the microvesicles was, furthermore, in accord with an external radius of 120 ± 4 Å and an internal radius of 73–86 Å.

The ultracentrifuge can be extremely useful in cooperation with X-ray, light-scattering, and other techniques in determining the shape and size of the vesicles. Using analytical ultracentrifugation, S. M. Johnson and Buttress (1972) have shown that measurable changes in size of vesicle and thickness of membrane occur on the addition of cholesterol.

Suspensions of multilamellar liposomes can be sedimented in a preparative ultracentrifuge to constant pellet volume in NaCl or KCl solution up to about 200 mM at speeds to 35000 rpm in 90 min. If higher molarities of salt are used, the lipids spin upward. Depending, too, on the proportion of ionizable amphiphilic molecules and on the ionic strength, the actual pellet volume can occupy up to eight times the molar volume of the lipid (see, for example, Fig. 5). The pellet thus formed is virtually a single phase, which is to say that every lipid bilayer is equidistant from its nearest neighbor. The extraliposome space can be estimated by adding ^{14}C-sucrose

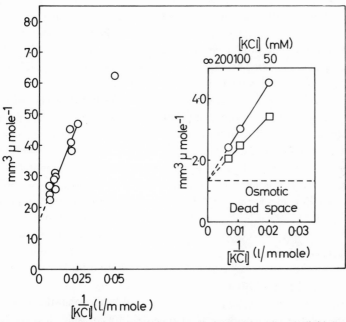

Fig. 9. **Osmotic swelling of phospholipid liquid crystals.** The phospholipid (composition as in Fig. 1) was allowed to form a dispersion in 145 mM KCl, and this was diluted with KCl solutions of various concentrations. The points represent pellet volumes (mm^3/μmole of phospholipid), at equilibrium, from a number of experiments plotted against the reciprocals of the KCl concentrations. Inset: Curves from two experiments, showing the conformity with the Boyle–van 't Hoff law. Here the pellet volumes have been corrected by subtraction of the interstitial volumes (10% of the pellet volumes). Extrapolation to infinite salt concentration gives the volume of the osmotic dead space. From Bangham et al. (1967b).

before centrifuging. The volume of this pellet will therefore depend on the intrinsic volume of the bilayer plus the free water space between bilayers. If, for instance, the osmotic pressure of the suspending medium is changed by adding a substance that is impermeable, the pellet formed on centrifugation will have a different volume, the difference being approximately the volume change of the sequestered water space. Bangham *et al.* (1967*b*) showed that this was also related to the optical density (Fig. 8). That the pellet volume is related to the osmolarity of the outside solution shows liposomes to act as osmometers to an appropriate solute and also to be permeable to water. Figure 9 shows how the suspension of lipid in fact obeys the Boyle–van't Hoff law and also gives an estimate of the osmotic dead space, from which an estimate of the amount of osmotically inactive water—about 25 water molecules per lipid—may be calculated. Figure 5, taken from the same paper, also has important application, for this shows the change in the lamellar spacing with changes in the salt concentration of the suspending medium. It is first noticed that pure lecithin (uncharged) is unaffected, but when phosphatidic acid is added the salt concentration has an appreciable effect on the pellet volume. This result gives some idea of the interaction between the lamellae and of the effect the electric double layer has in the equilibrium spacing. The situation is more complicated than was previously thought. Singer and Bangham (1971) have recently shown that when a smectic mesophase, particularly a charged one, is made permeable to both cation and anion, e.g., with I_2 and valinomycin, or when ammonium acetate is the prevailing salt they spontaneously imbibe salt and water until a new restraint to further swelling supervenes. It is likely that this restraint is an interfacial one.

3.6. Temperature Jump

Hammes and Tallman (1970) have described an experiment on the interaction of a dye with sonicated liposomes made from phosphatidylserine where the temperature of the suspension was changed rapidly from a low value to 25°C and the adsorption of the dye monitored spectrophotometrically. The apparatus is described by Erman and Hammes (1966). What is being measured is the time taken for the new equilibrium to be reached, and this characteristic time was shown to vary with sonication and concentration of phosphatidylserine. Pure, uncharged phosphatidylcholine was shown to have no relaxation, and it can be concluded that the charged-lipid–dye interaction is being observed. Adding calcium and osmotically shrinking the liposomes had no effect, although the addition of polylysine decreased the observed relaxation time considerably, also indicating a charge effect. The explanation, offered tentatively by the authors, of a conformation change of the lipids in the bilayer is possible, but there will be a change in bonding of the dye with temperature and it could be this that is being measured. More interesting, perhaps, is the possibility that the change has a direct bearing on the electrical double layer and of the rearrangement of cations (dye) in this region. This technique has not been fully exploited but might give much information on the structure and environment of lipid dispersions.

Owen *et al.* (1970) looked, by light scattering, at the simpler system of pure phosphatidylcholine microvesicles by temperature jump. They induced an 8 °C rise in temperature in 5 μsec and measured the relaxation times. No relaxations were observed with solid colloids such as gold sols or polystyrene spheres, but with the lipid vesicles a relaxation time of the order of milliseconds was observed. There seemed to be a correlation between the viscocity of the medium and the relaxation times and some indication that the relaxation could be due to the permeation of small molecules through the bilayer membrane.

3.7. Nuclear Magnetic Resonance

Certain nuclei which have magnetic moments will orientate themselves either parallel or antiparallel to the local magnetic field. The energy difference between the two configurations is small, and the energy required for a nucleus to change state can be readily applied by an oscillating magnetic field of the appropriate resonant frequency. When the oscillating field is removed, the extra energy of the nucleus is lost to other nuclei and to the general environment. The local magnetic field will depend explicitly on the environment of the nucleus and any applied field, and therefore the resonant frequency of the nucleus and its mode of decay give information of the local structure. The width of the absorption peak of energy is dependent on the range of local environments a nucleus finds itself in. If there is rapid molecular motion between these environments, this range will be effectively averaged out and hence the peak will narrow.

Three nuclei have been used to study the properties of lipid dispersions ^{1}H, ^{19}F, and ^{13}C. The proton is the most commonly used, but has the disadvantage that there are too many in an aqueous suspension. However, if the water is replaced by deuterium oxide, the protons from various parts of the lipid molecule can be studied separately, as they will be in different environments and exhibit chemical shifts, that is, shifts in resonant frequency. Penkett *et al.* (1968) give examples of the chemical shift of sonicated dispersions of egg yolk lecithin and phosphatidylserine (microvesicles) made up in D_2O. Penkett *et al.* (1968) also show that the intensity of the absorption of both the protons on the alkyl chain and the proton on the choline (for lecithin) increases with increasing time of sonication. This is stated to be due to the increased thermal motion of the small vesicles, causing a narrowing of the line width. There has been some doubt as to the interpretation, and it is possible that the lipids in sonicated vesicles are in a different "state" than those in the multilamellar smectic mesophase. When temperature effects are studied (Salsbury and Chapman, 1968), a change in the NMR spectra occurs that parallels the changes observed by X-ray and DSC, and is interpreted as a melting of the alkyl chains.

A more revealing method of determining the details of molecular motion in the vesicles has been used by Birdsall *et al.* (1971). This involves using a probe molecule, monofluorostearic acid, with the fluorine attached to various positions on the alkyl chain. The ^{19}F resonance is then observed and shows a different line width depending on the position of the substitution. This is consistent with the idea that the part of the chain near the terminal methyl group is the most mobile, while those near the head group region are hindered in their motion. Broad agreement was obtained by Hubbell and McConnell (1971) using spin-label techniques.

It was stated earlier that the protons from different regions could be looked at separately. This is indeed possible but involves a special technique to resolve the various line widths. This has been done for microvesicle dispersions of phosphatidylcholine by Lee *et al.* (1972), and they show marked changes in the spin-lattice relaxation time at the temperature of thermal transition. The resonance from the chains disappears and that from the choline protons increases in line width sharply as the chains solidify. The vesicle size remains constant through the transition, and the process is reversible. A similar sort of effect is observed on the addition of cholesterol to phosphatidylcholine microvesicles. This was first reported by Chapman and Penkett (1966) and repeated by Lee *et al.* (1972), and both authors suggest a large interaction between these lipids, the hydrocarbon region of the phosphatidylcholine being made much more "solid" by the presence of cholesterol.

The full potential of ^{13}C resonance spectra has yet to be realized, but it has been shown by Metcalfe *et al.* (1971) that high-resolution spectra can be obtained with smectic mesophases of dipalmitoyl phosphatidylcholine and that the carbon atoms from various parts of the molecule have different relaxation times depending on their distance from the glycerol moiety of the molecule. This is in contrast with the protons, which seem to have only one relaxation time (T_i) (Daycock *et al.*, 1971).

3.8. Spin Labels

Free radicals exhibit paramagnetic resonances, which as is the case with all resonant phenomena are dependent on local environment. There is a parallel with NMR, but with spin resonance we are looking at the change of spin state of an electron, not a nucleus. Normal chemical compounds have all electrons paired, and by the Fermi exclusion principle one must have the opposite spin to the other, so only where there is a free electron, i.e., a free radical, can resonances be obtained.

Most free radicals are unstable at room temperature, but certain organic ones have been found to be stable and can be given the name "spin label." The free radical normally used is the nitroxide group incorporated into a stabilizing ring. These molecules can be used to label parts of molecules by appropriate chemistry, and indeed this is one of the arts of the subject. A useful review of the physics and chemistry of spin labels is given by McConnell and McFarland (1970).

There are two methods. One is to start with a commercially obtainable free radical such as Tempo, 4-amino-22,66-tetramethylpiperidine (Aldrich Chemical Co.) and attach this via the amino group to whatever group is desired (Kornberg and McConnell, 1971). The other method is to use the reaction outlined by Jost *et al.* (1971), which follows accepted procedures. Hubbell and McConnell (1971) give methods for labeling fatty acids and lecithin; their reaction is

R_1 and R_2 are chosen to give the appropriate molecule and can be part of a ring, as with spin-labeled cholesterol. Resonance spectra were obtained by

Kornberg and McConnell (1971) using a commercially available instrument, the Varian E-4 spectrometer, and a sample size of 50 μl.

The information gathered by spin labels on the liquidity of the hydrocarbon chains agrees with that obtained by NMR and inferred from calorimetry (Hubbell and McConnell, 1971; Jost *et al.*, 1971). To simplify the theoretical interpretation, it is more convenient to have all the lamellae of the smectic mesophase parallel, and the samples are therefore built up on a glass surface and are not in the form of a vesicular suspension.

Kornberg and McConnell (1971) showed that it was possible to distinguish those spin-labeled molecules that were in the inner monolayer of a microvesicle from those that were in the outer monolayer. This was possible because the outside spin labels could be reduced by ascorbate at $0\,°C$, under which conditions ascorbate does not penetrate the vesicles. After a uniformly labeled suspension of vesicles was treated with ascorbate, and the excess ascorbate dialyzed away, the spectrum of only the inside spin labels could be measured. After a period of time, the spectrum was seen to revert back to the form it had before reduction. The conclusion was that the spin labels flip from the inside monolayer to the outside monolayer with a half time of 6.5 hr at $30\,°C$. Another experiment (Kornberg *et al.*, 1972), used a water-soluble spin label, tempotartrate, which is also permeable to the microvesicles. Again using ascorbate, the amount of spin label on the inside could be found, and hence the ratio of the anion concentrations, inside to outside, calculated. A potassium diffusion potential was set up using valinomycin, and the ratio of tartrate concentration was found to change. By the use of arguments similar to those used in deriving the Donnan equilibrium conditions, a predicted transmembrane potential of the order of 50 mV could be adduced. The ratios and potentials thus formed agreed with the known potassium ion concentration ratios.

3.9. Microelectrophoresis

In 1958, Bangham *et al.* described an electrophoretic apparatus which is most suitable for studying the surface charge properties of smectic mesophases. The design incorporates various features that are absolutely necessary if meaningful results are to be gained. For example, the whole of the sample tube and observing microscope objective are immersed in a temperature-controlled waterbath, and the positioning of the objective with respect to the sample chamber is rapid and accurate. The latter is of extreme importance, as there are *two* electrokinetic phenomena that occur in a closed glass tube containing an aqueous suspension of particles when an electric field is placed across the tube. There are the motion of the particle with respect to the surrounding medium due to the electric field and the motion of the liquid relative to the walls of the tube, the electroosmotic effect. There exists a position in a tube where this last effect is zero, the stationary layer, and if a cylindrical tube is used this can be readily calculated to be $0.293R$ from the wall of the tube of radius R. It is thus very important that the objective be of short focal length and focused exactly at this point. The cylindrical geometry does mean that the position of this theoretical stationary layer can be calculated and checked experimentally (Bangham *et al.*, 1958).

Electrophoresis measures the mobility W, which is the velocity of a particle per unit potential gradient. From this, the potential (ζ) at the plane of shear can be calculated using the following equation:

$$\zeta = \frac{4\pi\eta W}{D}$$

where η is the viscosity of the medium and D its dielectric constant.

In certain circumstances, it is possible to make the assumption that ζ is equal to the surface potential ψ_G and use the Gouy–Chapman equation to obtain the area per charge a where c is the univalent salt concentration:

$$\sinh\frac{\psi_G}{50.4} = \frac{134}{ac^{\frac{1}{2}}}$$

Haydon (1964) discusses in detail the validity of this assumption, and MacDonald and Bangham (1972) have shown that for low potential, less than 60 mV, this equivalence holds but above this the ζ potential is less than the predicated surface potential; that is, the plane of shear is significantly away from the plane of fixed charges and within the region of the counter ions. These latter experiments used various amounts of a charged species, phosphatidic acid, in phosphatidylcholine, not as a smectic mesophase, but as a stabilized emulsion of decane in salt solutions. Smectic mesophases of the pure lipids can be studied by this technique, and Figure 10 gives examples of three phospholipids and two mixtures showing the effect of changes as the bulk pH of the aqueous phase. It is perhaps important to notice that phosphatidylcholine has zero mobility down to pH 3, which makes it a useful neutral molecule for carrying other charged lipids into smectic mesophases and looking at the effects of pH on them (Bangham, 1961).

Two areas where electrophoretic data of smectic mesophases have been significant are blood clotting (Bangham, 1961) and the action of

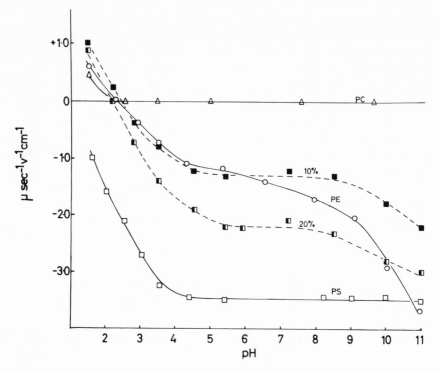

Fig. 10. Electrophoretic mobilities of various phospholipids as a function of *p*H. PC, Phosphatidylcholine; PE, phosphatidylethanolamine; PS, phosphatidylserine. Continuous curves are for pure compounds. Dashed curves are for mixtures of 10% and 20% phosphatidylserine in phosphatidylcholine. From Bangham (1968).

hydrolyzing enzymes on lipid mesophases (Bangham and Dawson, 1958, 1959, 1962); both reactions show strong correlation with the surface charge density.

4. METHODS

The most obvious parameter for measurement with liposomes is their permeability; numerous methods are available for such studies, and they may be applied to liposomes of constant composition with varying solutes, or *vice versa*.

Permeability of liposomes to solutes may be altered completely, e.g., following antigen–antibody–complement reactions (Kinsky, 1972), or incrementally, as with anesthetics (S. M. Johnson and Bangham, 1969a,b), or differentially to some ions, as with certain antibiotics (Henderson et al., 1969) or without (Papahadjopoulos, 1971). Permeability coefficients may vary from the exceedingly low values, e.g., 3×10^{-13} cm·sec^{-1} for some ions (S. M. Johnson and Bangham, 1969a) to a value for water of 1.0×10^{-4} cm·sec^{-1} (Bangham et al., 1967a). Naturally, methods have had to be developed and adapted to maximize accurate measurements over the very wide range. Furthermore, a claim for a coefficient implies a knowledge of the area across which solute is moving unidirectionally in unit time under a unit concentration gradient.

4.1. Surface Area

An estimate of the surface area of a homogeneous preparation of micro-vesicles of phospholipids may be derived separately or collectively from light-scattering, ultracentrifugation, diffusion, electron microscopic, spin resonance, NMR, and X-ray diffraction data, all of which techniques have been discussed earlier. The consensus is that a micromole of phosphatidyl-choline microvesicles, with or without a small charge component, presents an external surface area of approximately 2950 cm^2. This value is now supported by a direct method of measuring the surface area of a dispersion of liposomes making use of the high affinity constant of $UO_2{}^{2+}$ for phos-phate surfaces (Kruyt, 1948). In effect, a monolayer of the liposome phos-pholipid, being placed on a small volume (10 ml) of a salty (0.160 mM NaCl) solution, behaves, when appropriately connected, as a uranyl electrode (Fig. 11). Adsorption of $UO_2{}^{2+}$ from the bulk aqueous solution to the monolayer gives rise to a change ($\Delta\Delta V$) in the surface potential (ΔV) (Schulman and Rideal, 1931). The $\Delta\Delta V$ becomes a measure of the con-centration of free $UO_2{}^{2+}$ in the bulk phase. If, now, a relatively large area (compared to the monolayer) of adsorbing phospholipid, e.g., microvesicles, is added to the system, the free uranyl ion concentration falls and desorption from the monolayer takes place, resulting in a reduction in $\Delta\Delta V$. From the Gouy equation,

$$\Delta\Delta V = \psi_G = \frac{2\,kT}{e}\sinh^{-1}\left\{\frac{\sigma}{c_i^{\frac{1}{2}}}\left(\frac{500\pi}{DRT}\right)^{\frac{1}{2}}\right\} \qquad (1)$$

where c_i is the uni/univalent ion concentration in moles, k is the Boltzmann constant, T the absolute temperature, e the electronic charge, R the gas

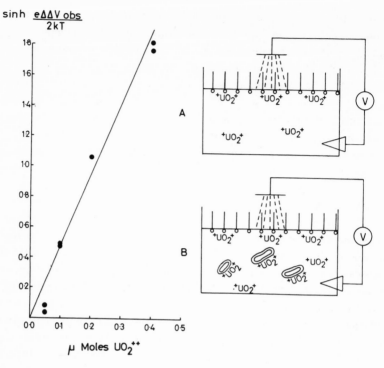

Fig. 11. Principles of the method and titration curve of liposomes. Full description of the method is given in Section 4.1.

constant, D the dielectric constant of water, and σ the surface charge density. Equation (1) reduces to

$$\sinh \frac{\Delta\Delta V}{50.54} = \frac{\sigma \times 2.8312 \times 10^{-5}}{0.16^{\frac{1}{2}}} \quad \text{at } 20\,^{\circ}\text{C} \tag{2}$$

if the surface potential and charge density are taken as zero, initially. If, too, the amount of UO_2^{2+} free in solution and on the monolayer is negligible in comparison with the amount on the large area of liposomes, e.g., the trough in Figure 10 is 10 cm² and 1 μmole of microvesicles has an area of approximately 3000 cm², then

$$\sigma = \frac{zeu \times 6.024 \times 10^{23}}{a} \tag{3}$$

where a is the surface area of liposomes, z is the number of charges added to the monolayer per UO_2^{2+} group, and u is the number of moles of uranyl-nitrate added.

When $z = 2$ and $e = 4.803 \times 10^{-10}$ esu, then

$$\sigma = \frac{u \times 5.787 \times 10^{14}}{a} \tag{4}$$

Substituting for σ in (2),

$$\sinh \frac{\Delta\Delta V}{50.54} = \frac{u \times 1.638 \times 10^{10}}{a\,(0.16)^{\frac{1}{2}}} \tag{5}$$

a can be calculated from the gradient of the graph sinh $(\Delta\Delta V/50.54)$ vs. u.

This method was developed and used for the measurement of the surface area of multilamellar (Bangham et al., 1967b) and microvesicle (S. M. Johnson et al., 1971) liposomes. The commonest cause of irreproducibility was a failure to remove all the UO_2^{2+} from the trough after a titration. This is because a monolayer of phosphatidylcholine adsorbs UO_2^{2+} significantly at UO_2^{2+} concentration of 5×10^{-7} M. If a minitrough of polytetrafluoroethylene is reused for a succession of measurements, it must be rinsed with a dilute solution (1 mM) of a divalent metal ion chelate such as ethylenediaminetetraacetate followed by meticulous acid and water rinses (MacDonald and Bangham, 1972). Alternatively, a fresh trough, which may be a small petri dish, immersed in molten thrice-recrystallized paraffin wax is used for each determination.

4.2. Dialysis

Most of the experiments that are carried out with liposomes involve a rate of leakage of a labeled or marker permeant from within the liposomes into a sink of known and larger volume. Since it is not practical to centrifuge liposomes out of solution as one would with biological cells or cell organelles, dialysis techniques are required and usually at two stages of an efflux experiment, namely, before and subsequently during the measuring period. The first dialysis is often referred to as "preparative" and amounts to a substitution of the untrapped label of the continuous aqueous phase by some isotonic, but unlabeled or different solute. Obviously, the advantages of measuring unidirectional fluxes at equilibrium as with isotopes are very attractive, and the preparative dialysis amounts to an exchange of an isotope for its chemically similar nonisotope. The choice of the dialysis technique depends, to a very large extent, on the permeability coefficient of the permeant under study. If, for example, the marked permeant has a very low permeability such as $^{22}Na^+$ or $^{42}K^+$, then the method described by Bangham *et al.* (1965a) has some merit. It involves pipetting aliquots (1.0 ml) of the liposome dispersion into lengths of wetted dialysis tubing tied off at their lower ends through a throat made from two concentric, snugly fitting lengths of polythene tubing which grips the dialysis tubing (Visking 21/32 inch). The inner polythene tube is then plugged. In batches of nine, these tubes are clipped to an all-polythene frame and placed into a 1-liter-capacity polythene bottle containing approximately 500 ml of isotope-free isotonic solution. The bottle and contents are then rotated at 1 rpm. From trial and error, it was found that efficient dialysis could only be attained when there is both thorough mixing of the dialysate—due to the tumbling action of the polythene frame—and gentle mixing of the dispersion—due to movement of air

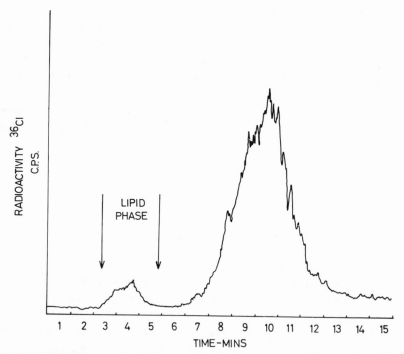

Fig. 12. A radioactivity profile of fractions emerging from a Sephadex column saturated with 0.145 M KNO₃. At zero time, 15 μmoles of a lecithin–5% dicetylphosphoric acid dispersion in 0.145 M ^{36}KCl was placed on the surface of the column. The lipids were washed through with KNO₃. When the Sephadex was saturated with Cl⁻ or I⁻ salts, no radioactivity was associated with the lipid fraction. From Bangham *et al.* (1965a).

bubbles inside the dialysis bag. It was found that five successive 30-min dialyses were necessary to reduce the concentration of untrapped marker to acceptably low levels.

Obviously, a $2^1/_2$-hr preparative dialysis might be an unacceptably long period for the more permeable solutes, i.e., those greater than 10^{-11} cm·sec^{-1}. It was fortunate, however, that a Pharmacia product, Sephadex G50, came on the market, for, at the suggestion of Papahadjopoulos et al. (1964), it was easily shown that liposomes of all sizes, even microvesicles, passed through these bead beds in the void volume, whereas ions and small solutes were retained (Bangham et al., 1965a,b). "Instant dialysis," as it became known, has proved to be an extremely useful preparative procedure (Fig. 12). Dimensions and load volumes of the bead beds are available from the literature (Papahadjopoulos and Watkins, 1967; S. M. Johnson and Bangham, 1969a; Klein et al., 1971). However, it is worth remarking on at least one important practical aspect of this procedure: namely, it is imperative to wash the polymer material very, very thoroughly, for, bearing in mind that the physical basis of smell is probably related to the penetration of olfactory membranes by (usually) volatile molecules, any residual contamination of the bead bed with soluble chemicals will modify the permeability properties of the liposomes in the same way as general anesthetics do (Bangham et al., 1965c; S. M. Johnson and Bangham, 1969b). As a more general warning of inadvertant contamination of aqueous systems by plastic containers, tubes, etc., the reader is referred to a recent cautionary tale by Bangham and Hill (1972).

4.3. Equilibrium Diffusion (Low Permeabilities)

Following the preparative dialysis, it is common practice to measure efflux rate constants by measuring the amount of the labeled permeant which emerges from the liposomes (a very small volume, 5–30 mm^3, but having a very large surface area) into a dialysis bag (much larger volume, 1–2 cm^3, but with a relatively small area for permeation) and finally into the sampling volume (usually about 10 cm^3).

The kinetic analysis of such a system is thus complicated both by the presence of the dialysis bag and, when using multilamellar liposomes, by the multiple and unknown number of compartments within the liposome. Setting aside the latter complication for the moment, errors caused by ignoring the presence of the dialysis bag can be important (Klein et al., 1971).

These systems can be described as quasi-equilibrium, where the analyses of the system reduce to considering the fluxes due to a concentration gradient. It is therefore important to define the quasi-equilibrium situation carefully and to see that the experimental situation falls within that definition. The condition for quasi-equilibrium is that across any interface the net flux of water, charge, and total solute is zero. The solute of interest is marked in some way so that it can be measured, either isotopically or by replacing it on one side of the system by a *close* analogue. The marker must have as near the same permeability as the analogue; otherwise, osmotic differences will be set up, causing water to move.

If the above conditions hold, we need only consider the movement of the isotope. The general condition will be as in Figure 13a, where we depict a series of m barriers to diffusion, the nth barrier having a permeability P_n, area A_n, and the volume contained between the barrier n and $n + 1$ equal to V_n. If the diffusion between barriers is much faster than across them, we can assume that the concentration there, C_n, is uniform, and hence the number of

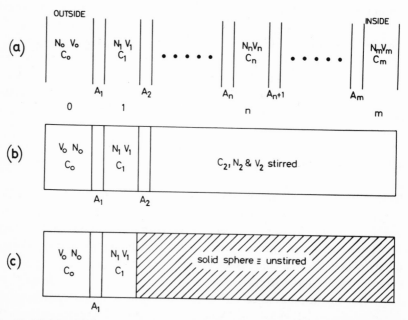

Fig. 13. **Models used in the analysis of the diffusion of marker from liposomes.** (a) Multilamellar system with m barriers to diffusion. (b) Two-barrier, three-stirred-compartment system where either the single-membraned microvesicles or a single membrane approximation to the multilamellar system is inside a dialysis bag. (c) Solid-sphere approximation inside a dialysis bag.

solute molecules $N_n = C_n \cdot V_n$. The volume V_0 is bounded by an impermeable membrane (the test tube), and it is the number of molecules here, N_0, that is the measured quantity. In general, each aqueous compartment is bounded by two barriers, and hence for the nth compartment

$$\frac{\partial N_n}{\partial t} = (C_{n+1} - C_n) A_{n+1} P_{n+1} - (C_n - C_{n-1}) A_n P_n \tag{6}$$

For the first compartment and last,

$$\frac{dN_0}{dt} = (C_1 - C_0) A_1 P_1 \tag{7}$$

$$\frac{dN_m}{dt} = (C_m - C_{m-1}) A_m P_m \tag{8}$$

The simplest case is of one barrier, say, a dialysis bag of area A_1 and volume V_1 and permeability P_1 containing $N_1{}^I$ counts at $t = 0$. If this is put into a volume V_0 containing no counts initially at time t,

$$\frac{dN_0}{dt} = \left(\frac{N_1}{V_1} - \frac{N_0}{V_0} \right) A_1 P_1 \tag{9}$$

$$\frac{dN_1}{dt} = - \left(\frac{N_1}{V_1} - \frac{N_0}{N_0} \right) A_1 P_1 \tag{10}$$

If we let $A_1 P_1 / V_1 = k_1$ and rearrange,

$$\frac{d^2 N_0}{dt^2} + k_1 \left(\frac{V_1}{V_0} + 1 \right) \frac{dN_0}{dt} = 0 \tag{11}$$

the solution of which is

$$N_0 = N_1^I \frac{V_0}{V_1 + V_0} \left\{ 1 - \exp - \left[\frac{k_1 t (V_0 + V_1)}{V_0} \right] \right\} \tag{12}$$

All terms in equation (12) are known except K_1 and, as will be seen later, the value of the dialysis bag constant is important and should in principle be determined for every marker used, and, as S. M. Johnson and Bangham (1969a) mention, it is also altered by several organic solvents and can vary from sample to sample.

The next simplest case for which an analysis has been attempted is for the single-membraned microvesicles (S. M. Johnson and Bangham, 1969a). These were contained in a dialysis bag so that altogether there were two barriers, k_1 for the bag and $k_2 = A_2 P_2 / V_2$ for the lipid bilayer of area A_2, permeability P_2, and trapped volume V_2.

Three equations have to be solved

$$\frac{dN_0}{dt} = k_1 N_1 - \frac{V_1}{V_0} N_0 \tag{13}$$

$$\frac{dN_1}{dt} = -k_1 N_1 - \frac{V_1}{V_0} N_0 + k_2 N_2 - \frac{V_2}{V_1} N_1 \tag{14}$$

$$\frac{dN_2}{dt} = -k_2 N_2 - \frac{V_2}{V_1} N_1 \tag{15}$$

The general solution in terms of N_0, the measurable quantity, is

$$N_0 = A \exp - \alpha t + B \exp - \beta t + C \tag{16}$$

If V_2/V_1 and $V_2/V_0 \ll 1$ then

$$\alpha = k_1 \left(1 + \frac{V_1}{V_0} \right)$$

and $\qquad\qquad\qquad\qquad\qquad\qquad\qquad\qquad\qquad\qquad\qquad$ (17)

$$\beta = k_2$$

The values of A, B, and C are determined by the conditions at $t = 0$. S. M. Johnson and Bangham (1969a) give details of these calculations and then make the following assumptions: First, as the permeability of the bag is orders of magnitude greater than the microvesicles, the bag can be ignored. Second, the fraction on the outside of the microvesicles but in the dialysis bag is small at the start of the experiment; this predicts a theoretical curve which is shown to hold over the early stages of the experiment up to about 60% of the release of the marker; after this, the amount released falls away sharply from the theoretical curve. This must undermine the confidence that can be placed in this analysis, and although the single microvesicle preparation looks at first sight theoretically easier to cope with than the multilamellar liposomes, it has not yet been analyzed with complete satisfaction, or alternatively the theoretical conditions cannot be met with in practice.

It is relevant at this point to note that any analysis of a unidirectional efflux out of a preparation of microvesicles is somewhat suspect for the simple reason that the number of solute molecules per microvesicle is only of the order of 10^2. If tracer atoms alone are being measured, few if any of the microvesicles would have more than two such marked atoms inside and a fraction will have none at all: normal diffusion kinetics would only apply if the system can be regarded as one vesicle with the area and internal volume of all the subunits of the system. Thus we assume that the time average of a single event is equal to the space average of that event over a large area (normal diffusion). This assumption (the ergodic hypothesis) has a long history of dispute and must warrant caution in the ultimate interpretation of permeability studies with microvesicles. An important point, however, which helps to confirm the validity of the diffusion process in microvesicles is the fact that the microvesicles exhibit different rate constants for K^+ with and without valinomycin but not for Na^+ or glucose, etc. Parenthetically, it is worth recalling that S. M. Johnson and Bangham (1969a) had to postulate, from their kinetic studies, that a valinomycin molecule does not reside indefinitely in a microvesicle but flits from one to another, ultimately releasing all of the trapped K^+. Kornberg et al. (1972) have recently confirmed this conclusion.

When considering the efflux profile of a multilamellar system, there is the possibility of trying to solve the system exactly or of making simplifying assumptions so as to make simple mathematical models. As a start, it is convenient, and in many cases justified, to neglect the effect of the dialysis bag. The first model to consider is that the outer lamellar is the only diffusion barrier (Fig. 13b) and to let the inside have an appropriate uniform concentration of marker permeant. Bangham et al. (1965a) found some experimental justification of this assumption, as the initial rates in their experiments followed the predicted linear relation with time. Chowhan et al. (1972) have looked at the model in detail by a computer fit and found that the details could not be reconciled to any permeability chosen. The next suggestion made (Bangham et al., 1965a) was that the system was like a hot solid sphere (Carslaw and Jaeger, 1947) losing heat into a stirred liquid (Fig. 13c). This would be a good model if the sphere were laminated with materials having alternating high and low thermoconductivity. Chowhan et al. (1972) used this approach and obtained good agreement between their model and experiment. The set of equations of the type (8–10) was analyzed on a computer, and moreover a full account was taken of the dialysis bag by measuring its rate constant first, without any liposomes present (see Fig. 14, from their paper). Chowhan assumed that when the liposomes were made (by solvent evaporation) the marker (^{14}C-glucose) was distributed between aqueous and lipid phases with a partition coefficient k and that there was no concentration gradient in any phase. The concentration in any aqueous compartment is related to the average concentration in any segment (that is,

44

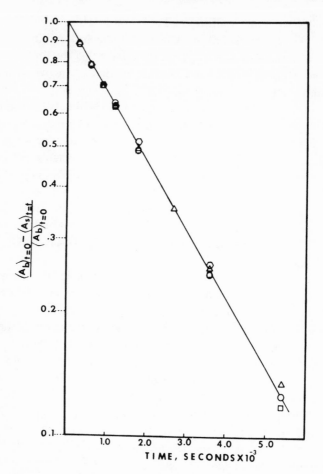

Fig. 14. First-order plots used for the determination of bag constant of D-glucose through a dialysis bag. The various symbols represent data from four experiments. From Chowhan *et al.* (1972).

the aqueous and one neighboring lipid phase) through the volume fraction of lipid and partition coefficients. The rate of change of the average number in a segment is then related to the concentrations in the aqueous compartments and the permeability and area of the interface. Two models were investigated, one counting the outer membrane as a separate entity from the rest and the other counting all membranes as equivalent. Both models gave satisfactory fits with the same permeabilities of 1.2×10^{-8} cm·sec^{-1} but

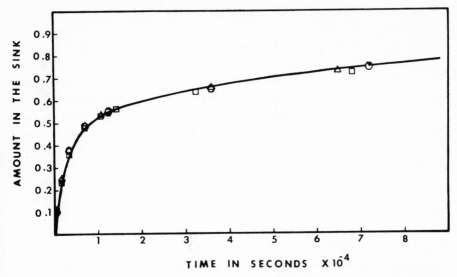

Fig. 15. Results from the dilution–release experiments showing the release of D-glucose as a function of time from a lecithin–dicetylphosphate dispersion system. Symbols represent experimental data. △, Dispersion system prepared from lecithin batch 1; □, dispersion system prepared from lecithin batch 2; ○, dispersion system 3, prepared from commercial lecithin; ⟨⟩ dispersion system 4, prepared from commercial lecithin. The curve represents theoretical calculations. From Chowhan *et al.* (1972).

slightly different values of the partition coefficients (Fig. 15). As the values of the lipid volume fractions and surface areas are not given in this paper, it is difficult to say if the high value of about 20 for the partition coefficient is as odd as it seems. The model also showed that as the number of bilayers increased a limiting value of curve slope was obtained at $n = 45$. When account was taken of the heterogeneity of the system, no change was seen with either model. In summary, these authors claim that two parameters, the partition coefficient and permeability, are all that is required to define any multilayered system, and the permeability thus measured is a good quantitative number to use in comparing different solutes or different lipid systems.

There are, it should be stated, dangers in concluding that the apparent efflux of a colored or isotopically labeled element or compound necessarily represents the permeability of what might be presumed to be the prevailing species in the solution. The reader is reminded, for example, that an element such as ^{36}Cl, represented in the form of a salt with K, *viz.* KCl, might exist in water as a number of species: Cl_2, KCl, HCl, Cl_3^-, Cl^-, of which perhaps only one may be freely permeable and reasonably abundant. With hindsight and fresh evidence (Singer and Bangham, 1971; MacDonald and Bangham, 1972), it is probable that the observed higher effluxes of ^{36}Cl than ^{42}K from negatively charged liposomes (Bangham, 1968; Papahadjopoulos and Watkins, 1967) were due to passage across the membrane of the uncharged species, HCl, and not the ion as casually assumed. Indeed, Kornberg et al. (1972) confirmed this interpretation rather elegantly by deliberately setting up a Cl$^-$ gradient and then measuring the equilibrium distribution of a spin label, tempotartrate, inside and outside a microvesicle preparation. On this argument, Sessa and Weissmann (1968), in following the release of the chrome chromophore, might well have been measuring the efflux of chromic acid! Since their chromate-loaded liposomes were dialyzed against isotonic KCl, it is further likely that HCl diffused down its concentration gradient into the liposome. Depending, therefore, on the relative abundance of HCl and H_2CrO_3, the liposomes should have tended either to swell or to shrink in response to the amount of solute moving into or out of the liposomes. There is also a further complication because HCl and H_2CrO_3 dissociate to yield an unequal number of protons, causing a pH shift and a further readjustment of all other colligative properties.

One still wonders how and in what disguise simple metal ions like Na$^+$ or K$^+$ actually do cross an undoped bimolecular membrane of phospholipids. Reassuringly, the situation is much simpler with nonelectrolytes, where it can rather simply be stated that either simple exchange of a labeled for an unlabeled molecule takes place or, if solute moves unidirectionally into or out of a liposome, water follows and a change in volume can be observed (Bangham et al., 1967b; Cohen and Bangham, 1972).

Methods are available which enable one to measure a permeant directly as it appears in or is lost from the continuous aqueous compartment, i.e., methods which eliminate the dialysis bag. Henderson et al. (1969) exploited earlier techniques, applied to mitochondria, which involved the placing of a small hydrogen and/or potassium selective glass electrode directly into the dispersion of liposomes. In this way, they were able to measure directly and quickly the appearance of potassium and disappearance of H$^+$ (Fig. 16). Scarpa and de Gier (1971) measured the same two ions but with the much more selective valinomycin-type K$^+$ electrode (Philips I.S. 560-K). It is advantageous and comparatively easy to measure changes in light scattering at the same time in these experiments, because one will not then overlook the unexpected net uptake of salt, as observed by Singer and Bangham (1971). Interestingly, valinomycin partitioned more favorably into

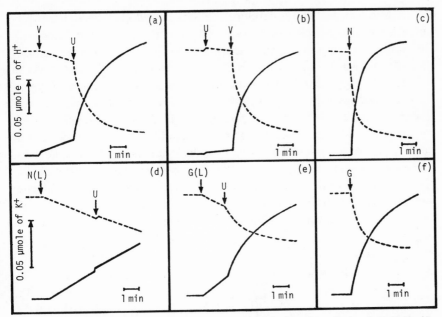

Fig. 16. Antibiotic-induced K⁺–H⁺ exchanges across the lamellae of phospholipid liquid crystals. Smectic mesophases of lecithin–10% dicetyl hydrogen phosphate containing 100 mM potassium mucate were prepared as described in the text, and external medium was exchanged for 145 mM choline chloride–0.5 mM KCl–0.75 mM-*N*-tris(hydroxymethyl)-methyl-2-aminoethanesulfonic acid, pH 6.65, on a Sephadex column. A 0.5-ml sample of the dispersion emerging from the column was suspended in 6.5 ml of the same medium at 30°C. The final lipid concentration was 0.23 μmole of lipid phosphate per milliliter. Additions were V, 0.5 μg valinomycin; U, 0.26 μM carbonylcyanide-*p*-trifluoromethoxy-phenylhydrazone (final concentration); G, 0.5 μg gramicidin D; N, 0.5 μg nigericin; G(L), 0.005 μg gramicidin D; N(L), 0.005 μg nigericin. In each case, the upper record (– – – –) indicates pH changes and the lower one (———) indicates changes in K⁺ concentration. The response of the electrodes was linear with respect to concentration of K⁺ and H⁺ over the ranges of H⁺ and K⁺ used. From Henderson *et al.* (1969).

the diphenyl ether of the electrode membrane than into the liposomes, a point to be watched for when using this type of electrode. As yet, commercially available assemblies which would permit the simultaneous monitoring of H⁺, K⁺, and light scattering, at constant temperature, are not available, and beginners in these fields are often at a loss to know whether to build around an expensive spectrophotometer or an expensive ion concentration–measuring device. No dogmatic advice is now offered, and the reader can only be advised to consult the relevant literature (Henderson *et al.*, 1969; Selwyn *et al.*, 1970; Scarpa and de Gier, 1971).

A technique that was developed by Kinsky *et al.* (1968) and extensively used by him during his studies with antigen-marked liposomes, lysed with antibody and complement, was the spectrophotometric absorbance changes at 340 mμ which occur when NADP is converted stoichiometrically to $NADPH_2$ by glucose (the permeant) by an appropriate mixture of enzymes. The method involves preparing a very modest amount of a suspension of multilamellar or microvesicle liposomes — 10 μmoles phospholipid/ml— made up in 300 mM glucose. After 2 hr equilibration, the suspension is "dialyzed" over Sephadex G50 or in a dialysis sac against isotonic KCl. Aliquots are then dispensed into separate tubes—0.2–0.3 ml is sufficient— and treated according to the nature of the experiment. When the glucose efflux is required to be measured, 5 μl only of the dispersion is added to a cuvette (1 cm light path) containing the following: 0.28 ml tris buffer (0.1 M,

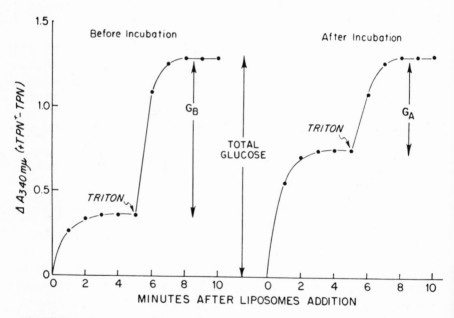

Fig. 17. Experiment to illustrate application of spectrophotometric assay. Liposomes (5 μl), prepared from egg lecithin and dicetylphosphate in a molar ratio of 7:2, were added to experimental ($+NADP^+$) and control ($-NADP^+$) cuvettes at zero time. Triton (0.1 ml), to give a final concentration of 1%, was added to both cuvettes at 5 min. The values on the ordinate are the differences in absorbance (corrected for dilution following each addition) between the experimental and control cuvettes. The curve on the left was obtained with liposomes assayed immediately after dialysis; that on the right was obtained after the liposomes had been incubated 1 hr at 41 °C. From Demel *et al.* (1968).

pH 8), 0.50 ml of 300 mM NaCl made up in tris buffer, 0.10 ml magnesium acetate (0.02 M), 0.05 ml ATP (0.02 M), 0.05 ml NADP (0.01 M), 5 μl hexokinase, and glucose-6-phosphate dehydrogenase. Control tubes are identical except that either NADP or glucose-6-phosphate dehydrogenase is omitted (Fig. 17). The method would be particularly useful when using microvesicles because it is at present not possible to follow the volume changes (if any) that would occur when glucose diffuses out into an isotonic salt solution. It also has some merit for assaying a membrane-active drug, because the aqueous compartment surrounding the liposomes need be very small, but reliance alone on the behavior of glucose permeation can be misleading (see, for example, Calissano and Bangham, 1971).

4.4. Nonequilibrium Fluxes (High Permeabilities)

If the liposome is to be accepted at all as a valid model of a biological cell or cell organelle membrane, it requires to be shown that it is some eight or nine orders (10^8–10^9) of magnitude more permeable to water than it is to sucrose or K$^+$. Fortunately, multilamellar liposomes, like mitochondria, are found to exhibit changes in light scattering such that their optical density [log (I_0/I), where I_0 is the intensity of the incident light] is linearly related to the reciprocal of their particle volume. Since changes in light intensity can be followed very fast indeed, it was only necessary to devise some rapid-mixing technique, ideally a fast-reaction apparatus, to exploit the movement (if any) of water by osmotic gradients. Bangham *et al.* (1967b) improvised such a method having only a moderate time resolution but from which they obtained a water permeability coefficient within an order of magnitude of the expected value. The apparatus and its manipulation are detailed below because fast-reaction apparatuses are not commonly available and yet, in principle, any sensitive spectrophotometer or simple white-light photometer may be adapted (Reeves and Dowben, 1970, use a stop-flow adaptor for use with a Zeiss PMQ II spectrophotometer). The adaptation was carried out on a Unicam SP800, a small d.c. motor driving a chemical stirrer; it was positioned so as to be outside the light beam and was run as fast as possible, cavitation and bubble formation being avoided. In order to avoid any loss of material through splashing, the mixed volume of liquid in the 1-cm cuvette was restricted to 2.5 ml. This restriction meant that the cuvette had to be raised on a shim and the height of the aperture in the holder reduced by a mask to 15 mm. After recording had started (chart speed 1 cm/sec), the dispersion of liposomes (0.2 ml) was injected in less than 0.5 sec from a Repette (Jencons

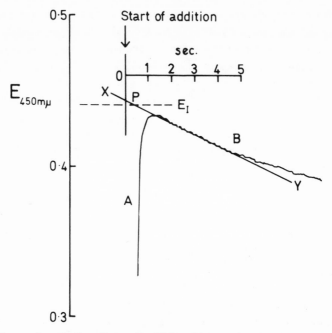

Fig. 18. **Changes in extinction after rapid addition of 0.2 ml of a dispersion of phospholipid (3.96 mM) formed in 40 mM KCl into 2.3 ml of water.** (See Section 4.4.) Composition of phospholipid as in Fig. 1. E_1 was the constant value of E when 0.2 ml of the dispersion was injected into 2.3 ml of 40 mM KCl. It was presumed that if the slope of the straight line XY (fitted to the first part of the swelling curve B) was to be a measure of the initial rate of change of E (dE/dt), then XY should intersect the line $E = E_1$ at some point (P) between the line $t = 0$ and the line A, which represents the rapid upward movement of the pen (Recordings in which this was not so were rejected.) Since $d/dt\,(1/E) = -1/E^2\,dE/dt$, the initial value of $d/dt\,(1/E)$ was obtained by multiplying the initial value of dE/dt by $-1/f_1^2$. From Bangham et al. (1967b).

Scientific Ltd.) through a fine polythene canula; this, inserted through a light-tight gland in the cell compartment cover, was guided by a glass tube into the fluid in the cuvette. For reproducible results, the cannula tip had to be in a fixed position near to the stirrer. A second channel of the recorder, actuated by a contact on the Repette, served as a time marker. Dispersions were pulsed alternatively into (1) the solution in which they were formed and (2) a solution of different concentration. Some idea of the scale expansions, liposome quantities, etc., may be obtained from Figure 18. De Gier et al. (1968) likewise have adapted a Vitatron UFD photometer to accommodate a double-walled cylindrical glass cuvette; with a fast-running stirrer in the cuvette, they claim complete mixing of a 50-μl sample of a multilamellar liposome suspension into 5 ml of isotonic medium in 1 sec or less. De Gier and colleagues have used this apparatus extensively for the study of composition–function relationships with both natural and artificial membranes (for review, see de Gier, 1972).

Since

$$\frac{d(1/E)}{dt} = -\frac{1}{E^2}\frac{dE}{dt} \tag{18}$$

the initial value of $d(1/E)/dt$ was obtained by multiplying the initial value of dE/dt by $-1/E_1^2$ (see Fig. 18). Values of $1/E$ must be converted to volumes from a calibration curve such as is illustrated in Figure 7. The osmotic water permeability coefficient k is given, in cm·sec^{-1} (if t is expressed in sec), by the equation

$$k = \frac{dv}{dt} \cdot \frac{1}{A\pi}\left[\frac{55.6 \times 22.4T}{273}\right] \cdot 10^4 \tag{19}$$

where v is the volume (in μ^3), A the area (in μ^2) measured by UO_2^{2+} titration, π the osmotic difference (in atmospheres), and T the absolute temperature. The quantities in the brackets convert to standard salt gradient; $\pi = RT(g_2 m_2 - g_1 m_1)$, where g and m are, respectively, the osmotic coefficients and molalities of the solute.

Where absolute measurements are not required, a measure of relative permeability to, say, an impermeant such as KCl is the reflection coefficient σ. These coefficients may be measured for various nonelectrolytes using the "zero-time" method of Goldstein and Soloman (1960). The method, again using light scattering and its empirical relationships between optical density and reciprocal volume, depend on determining a concentration of the permeant solute which initiates neither swelling nor shrinking at $t = 0$; i.e., the initial liposome volume change (dv/dt) t equals zero. Under these conditions,

$$\sigma_{\text{permeant}} = -\frac{\Delta C_{\text{impermeant}}}{\Delta C_{\text{permeant}}} \tag{20}$$

Lelievre has successfully used this approach in conjunction with a fast reaction (dead time 2 msec) apparatus, and his results support the idea that multilamellar liposomes can withstand the turbulence of a stop-flow operation.

Reflection coefficients, using "zero-time" kinetics, are not particularly

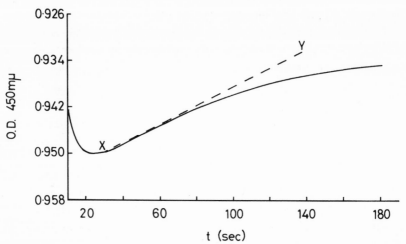

Fig. 19. Time course of the changes in absorbance. Changes were measured after rapid addition of 0.18 ml (8 mM lipid concentration) suspension of liposomes formed in 20 mM KCl into 2.5 ml of 100 mM ethylurea at 10 °C. Liposomes were prepared from a mixture of egg yolk lecithin, cholesterol, and phosphatidic acid in a molar ratio 48:48:4, respectively. Printout readings of the absorbance changes were taken at least every second. From a sequence of readings taken after the minimum, a maximum slope was calculated. The portion of the straight line XY common to the absorbance curve illustrates the extent of the sequence that can be used for calculating the slope for ethylurea. From Cohen and Bangham (1972).

discriminatory because solutes such as urea, glycerol, and glucose all give values of 1 with respect to KCl.

An alternative and quite sensitive method for measuring the relative permeability of solutes whose permeability is of the order 10^{-6} cm·sec^{-1} was one devised theoretically many years ago by Jacobs (1933), and it has recently been applied with some success to multilamellar liposomes by Cohen and Bangham (1972). These authors, using cholesterol-containing liposomes mixed rapidly with hypertonic solutions of solutes at 10 °C, found that the osmotic movements of water and solute could be so slowed that two stages could be resolved (Fig. 19). First, an initial slope after mixing (complete within 200 msec) represents an increase in optical density indicative of a shrinking due to the high outside–inside solute gradient. This slope terminates at a minimum-volume equilibrium point whose depth and time after mixing are related to the permeability coefficient of the external solute. Thereafter, there is a progressive decrease in the optical density representing an influx of external solute (down its concentration gradient) and its concomitant water. The time that the system takes to achieve the minimum volume, or the derivative of the rate of volume change at that minimum, has been used by others to determine the permeability coefficient of the permeant. Two equations describe the events:

$$\frac{dn_s}{dt} = P_s A \left(C_s - \frac{n_s}{V} \right) \tag{21}$$

$$\frac{dV}{dt} = P_w A \left[\left(\frac{n_i}{V} - C_1 \right) + \sigma \left(\frac{n_s}{V} - C_s \right) \right] \tag{22}$$

where C_s is the concentration of permeable solute outside the liposomes, C_1 is the concentration of impermeable solute outside the liposomes, n_s is the number of permeable solute molecules in the liposomes, n_i is the number of impermeable solute molecules in the liposomes, A is the area of liposomes, V is the volume of the liposomes, P_s and P_w are the solute and water permeability coefficients, respectively, and σ is the reflection coefficient of permeable solute.

These equations were solved numerically by hand in 1933 by Jacobs, putting $\sigma = 1$ and values for various experimental parameters. Fortunately, information can be extracted from experimental data without recourse to solving the equation in this manner. If the liposomes are made up in 20 mM KCl, for example, which is impermeable, and then 100 mM of permeable solute is added to the outside, the liposomes first shrink under the osmotic force and then swell due to movement of permeable solute (see Fig. 19). We can make the assumption that at the minimum volume the osmotic pressure across the liposome is zero and will remain zero from that time on, as any solute that moves will be accompanied by a net movement of water. The argument assumes water to be very much more permeable than the solute, which will normally be the case. The condition for minimum volume is that

$$\frac{dV}{dt} = 0$$

$$\frac{n_1 + \sigma n_s}{V} = C_1 + \sigma C_s \tag{23}$$

After the minimum volume, we find experimentally that there is always a region where the volume changes with time in a linear fashion; this means that $d^2 V/dt^2 = 0$.

Differentiating equation (22), we get

$$\frac{-n_1}{V_2}\frac{dV}{dt} + \frac{1}{V}\frac{dn_1}{dt} - \frac{dC_1}{dt} - \frac{\sigma n_s}{V^2}\frac{dV}{dt} + \frac{\sigma}{V}\frac{dn_s}{dt} - \frac{\sigma dC_s}{dt} = 0 \tag{24}$$

We can assume that the concentration of the external solution remains constant, as the volume trapped in the liposomes is orders of magnitude smaller than the outside, so dC_1/dt and dC_s/dt both equal zero. Also, the number of impermeant molecules inside does not change, so $dN_1/dt = 0$.

Equation (21) gives dN_s/dt, so we get

$$\frac{dV}{dt} = P_s A \sigma \frac{C_s - N_s/V_\sigma}{N_1 + N_\sigma} \tag{25}$$

Using the osmotic equilibrium condition equation,

$$\frac{dV}{dt} = \sigma P_s A \frac{C_s - N_s/V}{C_1 + \sigma C_s} \tag{26}$$

If the osmotic pressure of the permeable solute outside is much larger than the impermeant, then $\sigma C_s \gg C_1$. Also, if up to the minimum volume mostly water moves, the concentration of permeable solute inside will be small compared to that outside, so $C_s \gg N_s/V$; hence,

$$\frac{dV}{dt} = P_s A \tag{27}$$

Thus the slope of the volume change after the minimum volume, if linear, will be proportional to the permeability of the permeable solute. We can record the volume change by measuring the extinction of the solutes and use the empirical relation $V \alpha (1/E)$:

$$\frac{dV}{dt} = \frac{-1}{E^2}\frac{dE}{dt} \tag{28}$$

If one sample of liposomes is used to compare many solutes, the value of E will be the same for all the solutes initially, and, as the changes observed are small, relative values of dE/dt will be proportional to relative values of dV/dt or the permeability.

Cohen and Bangham (1972) have used this method to compare the diffusion of various nonelectrolytes across liposomes. Hill and Cohen (1972) gave further details of this analysis. J. A. Johnson and Wilson (1967), Sha'afi et al. (1970), and Sigler and Janáček (1971) have used other methods of analyzing the osmotic changes of natural membranes, and details of their analysis can be found in their papers.

5. MATERIALS

Phospholipids were first observed in 1811 in material isolated from brain. Gobley, in 1850, proposed the name lecithin ("lecithine"), and Strecker (1868) gave lecithin its correct formula. Later, Thudichum (1874) achieved a partial separation of the phospholipids and was the first to isolate sphingomyelin. He also introduced some of the early terminology. Folch (1942) made a major breakthrough in phospholipid chemistry when he isolated a fraction (from brain tissue) which he called "cephalin." This could be further separated into three fractions: phosphatidylinositol, phosphatidylserine, and phosphatidylethanolamine. Subsequently, column chromatography of phospholipids was reported by Thannhauser and Setz (1936), Hanahan et al. (1951, 1957), Lea et al. (1955), and Dawson (1958). Bligh and Dyer (1959), studying phase relationships of chloroform–methanol–water, presented a method whereby the lipids of biological membranes can be extracted and purified in a single operation.

A simple biochemical method for the preparation of high specific activity ^{14}C-acyl-labeled phosphatidylcholine and other phospholipids is given by Galliard (1972), using the useful and practical extraction procedure of Bligh and Dyer (1959).

Several books (Hanahan, 1960; Ansell and Hawthorne, 1964; Marinetti, 1967; A. R. Johnson and Davenport, 1971) and journals *(B.B.A. Lipids, J. Lipid Res., J. Am. Oil Chem. Soc., Lipids, J. Biol. Chem., J. Chromatog., Biochem. J., etc.)* are available for a further study on methods.

5.1. Egg Phosphatidylcholine

Since liposome systems demand rather large quantities of phospholipids, we have developed methods in this laboratory for preparing phosphatidylcholine, and subsequently phosphatidic acid, using, but modifying to some extent, existing techniques (Hanahan et al., 1951, 1957; Lea et al., 1955; Dawson, 1958; Papahadjopoulos and Miller, 1967).

Phosphatidylcholine (egg lecithin; 1,2-diacyl-3-sn-o-glycerylphosphorylcholine), a mixed species of varying unsaturation, is one of the commonest lipids found and is also a ubiquitous lipid of membranes.

Rhodes and Lea (1957) give the composition of hen's egg yolk phospholipids in moles % as phosphatidylcholine, 73.0; lysophosphatidylcholine, 5.8; phosphatidylethanolamine, 15.0; lysophosphatidylethanolamine, 2.1; inositol phospholipid, 0.6; sphingomyelin, 2.5; and plasmalogen, 1.0.

5.1.1. Extraction and Purification

All extractions are carried out at 15–25 °C using analytical grade reagents where possible. The alumina and silicic acid are washed, after activation at 105 °C, in the appropriate solvents. Washing, prior to packing, removes "fines" that tend to block the sinter of the glass column. Six lyophilized egg yolks are homogenized with 150 ml acetone in a top-drive macerator, allowed to stand for 15 min, then centrifuged at 400 × g for 15 min to remove acetone-soluble impurities (triglyceride fat, steroids, and pigments). This is repeated four or five times or until the supernatant is quite colorless. The lipids are extracted from the residue with 500 ml chloroform–methanol (1:1, v/v), for $^1/_2$ hr at room temperature, centrifuged, and the supernatant is taken to a damp mass on a suitable rotary evaporator ("Rotavapor", W. Büchi, Flawil, Switzerland). The residue is dissolved in 50 ml petroleum ether (60–80 °C) and precipitated with 400 ml acetone. Finally, all traces of solvent are removed from the crude egg phospholipid, which is weighed and stored in a minimal volume of chloroform–methanol (1:1, v/v). Ten grams of the crude egg phospholipids is passed through a column of 320 g of well-washed alumina (aluminium oxide 100–240 mesh, Brockmann activity 1.0. Alkaline. CAMAG. M.F.C., Hopkin and Williams, diameter to height ratio, 1:10) using chloroform–methanol (1:1, v/v). Phosphatidylethanolamine and lysophosphatidylethanolamine are retained by the column, and crude phosphatidylcholine is eluted with the solvent front in about 500 ml. The lipid eluant is taken to dryness, weighed, dissolved in a minimum of chloroform–methanol (2:1, v/v), and stored under N_2 at −20 °C until required. Enough silicic acid (Silicar CC4 100–200 mesh or silicic acid 100 mesh, Mallinckrodt) is activated at 105 °C for the final column so that 1 g of phosphotidylcholine is separated on 40 g of silicic acid. With the column packed (diameter–height, 1:20) and chloroform–methanol (2:1, v/v) 2–3 mm above the silicic acid, the crude phosphatidylcholine (plus solvent) is added to the top of the column. The stopcock is opened to allow the lipid to soak in to within 1 mm and closed. Twenty to thirty milliliters of chloroform–methanol (2:1, v/v) is added slowly down the sides of the column, and the stopcock is opened again. This is repeated four times to wash all the lipid into the silicic acid. Finally, enough chloroform–methanol (2:1, v/v) is added to fill the column, and the stopcock is opened. If 100–200 mesh silicic acid is used, we find it necessary to apply gentle nitrogen pressure (2–3 lb/in.^{2}) to increase flow rate. Free fatty acids and some pigment (residual from an earlier cleaning attempt) are eluted at about 1 column volume. Pure phosphatidyl-

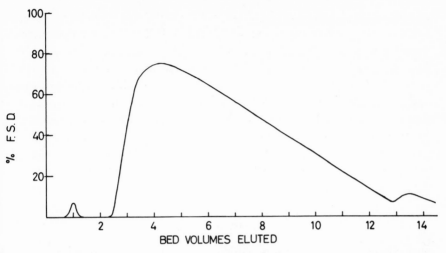

Fig. 20. Monitoring of the column eluent. This was done either by using a Pye liquid chromatograph with argon ionization detector or by analyzing every fourth tube for phosphorus (after evaporating solvent and digesting with 72% v/v perchloric acid) (Fiske and Subbarow (1925). Free fatty acids and residual pigment eluted at about 1 column volume, phosphatidylcholine at 2–3 column volumes, and lysophosphatidylcholine at 16–20 column volumes.

**Table I. Fatty Acid Methyl Esters of
Egg Phosphatidylcholine[a]**

Fatty acid	Weight % fatty acid
16:0	27.0
16:1	1.70
18:0	13.7
18:1	24.0
18:2	11.7
20:4	4.0
22:4, 22:5	1.5
22:6	5.0

[a] Average of five preparations.

choline is eluted at 2–3 column volumes (Fig. 20), the more polyunsaturated species first (Rhodes and Lea, 1956; Klein, 1970b). The solvent is collected until lysophosphatidylcholine appears, and all appropriate fractions are pooled and taken to dryness. Chloroform or ethanol is added, and the solution is flushed with nitrogen and stored at $-20\,°C$. As a rough guide, English eggs yield 0.7 g phosphatidylcholine per yolk. The methyl esters of the fatty acids of the phosphatidylcholine are prepared by transmethylation in methanolic sodium hydroxide (Dawson, 1954; Hübscher *et al.*, 1960) and chromatographed on a Pye series 104 gas chromatograph (see Table I). Other extraction methods have been extensively reported, e.g., using cadmium chloride precipitation (Pangborn, 1951), chromatography on alumina (Hanahan *et al.*, 1951), alumina followed by silicic acid (Lea *et al.*, 1955), and careful column loading and correct solvent ratios on alumina alone (Singleton *et al.*, 1965).

5.2. Phosphatidic Acid

Phosphatidic acid can be prepared from phosphatidylcholine by enzymic hydrolysis (Papahadjopoulos and Miller, 1967). The enzyme, phospholipase D, is prepared from either the inner white leaves of a Savoy cabbage or the white stalks of celery. Typical yields of enzymatic activity from the two starting materials are cabbage 30 and celery 80 μg choline liberated/min/mg dry weight of starting material (Davidson and Long, 1958; Quarles and Dawson, 1969). Two hundred grams of plant tissue is homogenized with 500 ml H_2O for 5 min. The fibrous material is filtered using "butter muslin" (cheesecloth), and the filtrate is centrifuged at 1300 $\times$ g for 30 min in a Sorvall centrifuge (Ivan Sorvall, Norwalk, Conn.). The crude enzyme preparation (decanted supernatant), stored frozen, is stable for many months.

Briefly, 5 g of dry phosphatidylcholine is dispersed in 500 ml 0.1 M acetate buffer (pH 5.6) and mixed with 350 ml of the crude enzyme solution, 100 ml 1 M $CaCl_2$, and 400 ml diethyl ether. The mixture is agitated for 6 hr at $25\,°C$ and then allowed to stand for the phases to separate. The Ca^{2+} salt of the phosphatidic acid is sandwiched between a clear upper (ether) phase and a lower (aqueous) phase. The white middle phase is converted to the acid, sodium, or potassium form by the method of Folch *et al.* (1957). After washing, the phosphatidic acid is purified by passing through silicic acid and eluted with chloroform–methanol (96:4, v/v). The final product is identified by thin-layer chromatography and stored at $-20\,°C$ under nitrogen. Higher yields of saturated phosphatidic acids, e.g., dipalmitoyl, are claimed by a method described by Kornberg and McConnell (1971), but the barium–phosphatidic acid complex can be dissociated using hydrochloric acid rather than sulfuric acid (Klein, 1972).

5.3. Ox-Brain Phospholipids

Brains obtained immediately after slaughter are processed to phosphatidyl-serine, phosphatidylethanolamine, phosphatidylinositol, cerebrosides, or sulfatides using the technique described by Folch (1942) and Papahadjopoulos and Miller (1967). Further fractionation may be performed on DE-32 cellulose (Whatman) following the method described by Rouser *et al.* (1961, 1963). Gangliosides are prepared by the technique of Penick *et al.* (1966) and sulfatides by the method of Lees *et al.* (1959).

5.4. Characterization and Purity

All phospholipids prepared by the above method should be examined by two-dimensional thin-layer chromatography (Abramson and Blecher, 1964) or by using an "S" chamber (Parker and Peterson, 1965) with a two-solvent system. The latter, fitted around the thin-layer plate, improves the compactness of the spots and decreases the development time.

New batches of phosphatidylcholine and phosphatidic acid should be routinely examined on activated silica gel H plates (0.25 mm thick) using chloroform–methanol–7 M aqueous ammonia (230:90:15, v/v/v) (Abramson and Blecher, 1964) and activated silica gel G with chloroform–methanol–water (65:25:4, v/v/v). The lipid solution is usually applied 3 cm from the edge of the plate by touching a calibrated glass capillary or microsyringe onto the layer without disturbing it. Loads of 10–500 μg lipid per spot are used. Compounds that run close together are better separated by lower loading. When minor components (contaminants) are being sought, higher loadings are necessary.

Saturated phosphatidylcholines can be completely separated from unsaturated phosphatidylcholines by thin-layer chromatography on silica gel plates impregnated with silver nitrate (Arvidson, 1965). A review by Mangold (1969) on "aliphatic lipids" gives many other methods for lipid separation.

5.4.1. Simple Specific Sprays for Lipids

For all organic compounds, plates are sprayed with 0.005 % rhodamine 6G and viewed under ultraviolet light or sprayed with 50 % H_2SO_4 and charred at 110 °C for 16 min. Only the rhodamine may be followed by other sprays. For lipids with a free amino group (e.g., phosphatidylserine and phosphatidyl-ethanolamine), plates are sprayed with 0.25 % ninhydrin in acetone and heated at 100 °C for 5 min. Finally, all phosphorus-containing lipids may be visualized by spraying with the molybdenum blue reagent (Dittmer and Lester, 1964).

5.4.2. Other Assays

Fatty acyl chains may be methylated and identified on a Pye series 104 gas liquid chromatograph (e.g., Table I). The fatty acids-to-phosphorus and phosphorus-to-nitrogen ratios can therefore be worked out. Oxidation of the lipids may be examined by the method of Klein (1970a). Sphingomyelin and phosphatidylcholine may be examined for purity by microelectrophoresis (Bangham et al., 1958). Finally, the purity of the isolated phospholipids may be checked by the successive hydrolysis technique of Dawson et al. (1962).

ACKNOWLEDGMENTS

We thank Mrs. Janet Hood for secretarial assistance and the University of Sheffield, Biomedical Information Project, for their discerning and prompt selection of titles and references in the field of cell and model membranes.

6. REFERENCES

Abramson, D., and Blecher, M., 1964, Quantitative two-dimensional thin-layer chromatography of naturally occurring phospholipids, J. Lipid Res. 5:628.
Abramson, M. B., Katzman, R., and Gregor, H. P., 1964a, Aqueous dispersions of phosphatidylserine ionic properties, J. Biol. Chem. 239:70.
Abramson, M. B., Katzman, R., Wilson, C. E., and Gregor, H. P., 1964b, Ionic properties of aqueous dispersions of phosphatidic acid, J. Biol. Chem. 239:4066.
Ansell, G. B., and Hawthorne, J. N., 1964, "Phospholipids," Elsevier, Amsterdam.
Arvidson, G. A. E., 1965, Fractionation of naturally occurring lecithins according to degree of unsaturation by thin-layer chromatography, J. Lipid Res. 6:574.
Attwood, D., and Saunders, L., 1965, A light scattering study of ultrasonically irradiated lecithin sols, Biochim. Biophys. Acta 98:344.
Bangham, A. D., 1961, A correlation between surface charge and coagulant action of phospholipids, Nature 192:1197.
Bangham, A. D., 1963, Physical structure and behaviour of lipids and lipid enzymes, Advan. Lipid Res. 1:65.
Bangham, A. D., 1968, Membrane models with phospholipids, Prog. Biophys. Mol. Biol. 18:29.
Bangham, A. D., 1972, Lipid bilayers and biomembranes, Ann. Rev. Biochem. 41: 753.
Bangham, A. D., and Dawson, R. M. C., 1958, Control of lecithinase activity by the electrophoretic charge on its substrate surface, Nature 182: 1292.

Bangham, A. D., and Dawson, R. M. C., 1959, The relation between the activity of a lecithinase and the electrophoretic charge of the substrate, *Biochem. J.* **72**: 486.

Bangham, A. D., and Dawson, R. M. C., 1962, Electrokinetic requirements for the reaction between *Cl. perfringens* α-toxin and phospholipid substrates, *Biochim. Biophys. Acta* **59**:103.

Bangham, A. D., and Hill, M. W., 1972, Distillation and storage of water, *Nature* **237**:408.

Bangham, A. D., and Horne, R. W., 1964, Negative staining of phospholipids and their structural modification by surface-active agents as observed in the electron microscope, *J. Mol. Biol.* **8**:660.

Bangham, A. D., Flemans, R., Heard, D. H., and Seaman, G. V. F., 1958, An apparatus for micro-electrophoresis of small particles, *Nature* **182**:642.

Bangham, A. D., Standish, M. M., and Watkins, J. C., 1965a, Diffusion of univalent ions across the lamellae of swollen phospholipids, *J. Mol. Biol.* **13**:238.

Bangham, A. D., Standish, M. M., and Weissmann, G., 1965b, The action of steroids and streptolysin S on the permeability of phospholipid structures to cation, *J. Mol. Biol.* **13**:253.

Bangham, A. D., Standish, M. M., and Miller, N., 1965c, Cation permeability of phospholipid model membranes: Effect of narcotics, *Nature* **208**:1295.

Bangham, A. D., Standish, M. M., Watkins, J. C., and Weissmann, G., 1967a, The diffusion of ions from a phospholipid model membrane system, *Protoplasma* **63**:183.

Bangham, A. D., de Gier, J., and Greville, G. D., 1967b, Osmotic properties and water permeability of phospholipid liquid crystals, *Chem. Phys. Lipids* **1**:225.

Birdsall, N. J. M., Lee, A. G., Levine, Y. K., and Metcalfe, J. C., 1971, ^{19}F NMR of monofluorostearic acids in lecithin vesicles, *Biochim. Biophys. Acta* **241**:693.

Bligh, E. G., and Dyer, W. J., 1959, A rapid method of total lipid extraction and purification, *Can. J. Biochem. Physiol.* **37**:911.

Bonting, S. L., and Bangham, A. D., 1967, On the biochemical mechanism of the visual process, *Exptl. Eye Res.* **6**:400.

Calissano, P., and Bangham, A. D., 1971, Effect of two specific proteins (S100 and 14.3.2) on cation diffusion across artificial lipid membranes, *Biochem. Biophys. Res. Commun.* **43(3)**:504.

Carslaw, H. S., and Jaeger, J. C., 1947, "Conduction of Heat in Solids," Oxford University Press, London.

Chapman, D., and Fast, P. G., 1968, Studies of chlorophyll–lipid–water systems, *Science* **160**:188.

Chapman, D., and Penkett, S. A., 1966, Nuclear magnetic resonance spectroscopic studies of the interaction of phospholipids with cholesterol, *Nature* **211**:1304.

Cherry, R. J., and Chapman, D., 1969, Optical properties of black lecithin films, *J. Mol. Biol.* **40**:19.

Chowhan, Z. T., Yotsuyanagi, T., and Higuchi, W., 1972, Model transport studies utilizing lecithin spherules. 1. Critical evaluation of several physical models in the determination of the permeability coefficient, of glucose, *Biochim. Biophys. Acta* **266**:320.

Cohen, E., and Bangham, A. D., 1972, Diffusion of small nonelectrolytes across liposome membranes, *Nature* **236**:173.

Danforth, R., Krakaver, H., and Sturtevant, J. M., 1967, Differential calorimetry of thermally induced processes in solution, *Rev. Sci. Inst.* **38**:484.

62

Davidson, F. M., and Long, C., 1958, The structure of the naturally occurring phospho-glycerides. 4. Action of cabbage leaf phospholipase D on ovolecithin and related substances, *Biochem. J.* **69:**458.

Dawson, R. M. C., 1954, The measurement of ^{32}P labelling of individual kephalins and lecithin in a small sample of tissue, *Biochim. Biophys. Acta* **14:**374.

Dawson, R. M. C., 1958, Studies on the hydrolysis of lecithin by a *Penicillium notatum* phospholipase B preparation, *Biochem. J.* **70:**559.

Dawson, R. M. C., Hemington, N., and Davenport, J. B., 1962, Improvements in the method of determining individual phospholipids in a complex mixture by successive chemical hydrolyses, *Biochem. J.* **84:**497.

Daycock, J. T., Darke, A., and Chapman, D., 1971, Nuclear relaxation (T_1) measurements of lecithin water systems, *Chem. Phys. Lipids* **6:**205.

Deamer, D. W., Leonard, R., Tardieu, A., and Branton, D., 1970, Lamellar and hexagonal lipid phases visualized by freeze-etching, *Biochim. Biophys. Acta* **219:**47.

de Gier, J., Haest, C. W. M., van der Neut-Kok, E. C. M., Mandersloot, J. G. and van Deenen, L. L. M., 1972, Correlations between liposomes and biological membranes. pp.263–278, Eighth FEBS Meeting, Vol. 28, *Mitochondria and Biomembranes*, North-Holland.

de Gier, J., Manderslot, J. G., and van Deenen, L. L. M., 1968, Lipid composition and permeability of liposomes, *Biochim. Biophys. Acta* **150:**666.

de Gier, J., Manderslot, J. G., and van Deenen, L. L. M., 1969, The role of cholesterol in lipid membranes, *Biochim. Biophys. Acta* **173:**143.

de Gier, J., Manderslot, J. G., Hupkes, J. V., McElhaney, R. N., and van Beek, W. P., 1971, On the mechanism of nonelectrolyte permeation through lipid bilayers and through biomembranes, *Biochim. Biophys. Acta* **233:**610.

Demel, R. A., Kinsky, S. C., Kinsky, C. B., and van Deenen, L. L. M., 1968, Effects of temperature and cholesterol on the glucose permeability of liposomes prepared with natural and synthetic lecithins, *Biochim. Biophys. Acta* **150:**655.

Dervichian, D. G., 1964, The physical chemistry of phospholipids, *Prog. Biophys. Mol. Biol.* **14:**265.

Dittmer, J., and Lester, R. C., 1964, A simple specific spray for the detection of phospholipids on thin-layer chromatograms, *J. Lipid Res.* **5:**126.

Erman, J. E., and Hammes, G. G., 1966, Versatile stopped-flow temperature jump apparatus, *Rev. Sci. Inst.* **37:**746.

Finean, J. B., and Millington, P. F., 1955, Low angle X-ray diffraction study of the polymorphic forms of synthetic $\alpha{:}\beta$ and $\alpha{:}\alpha'$-kephalins and $\alpha{:}\beta$-lecithins, *Trans. Faraday Soc.* **51:**1008.

Finer, E. G., Flook, A. G., and Hauser, H., 1972, Mechanism of sonication of aqueous egg yolk lecithin dispersion and nature of the resultant particles, *Biochim. Biophys. Acta* **260:**49.

Fiske, C. H., and SubbaRow, Y., 1925, The colorimetric determination of phosphorus, *J. Biol. Chem.* **66:**375.

Fluck, D. J., Henson, A. F., and Chapman, D., 1969, Structure of dilute lecithin–water systems revealed by freeze-etching and electron microscopy, *J. Ultrastruct. Res.* **29:**416.

Folch, J., 1942, Brain cephalin, a mixture of phosphatides. Separation from it of phosphatidylserine, phosphatidylethanolamine and a fraction containing an inositol phosphatide, *J. Biol. Chem.* **146:**35.

Folch, J., Lees, M., and Sloane-Stanley, G. H., 1957, A simple method for the isolation and purification of total lipids from animal tissues, *J. Biol. Chem.* **226**:497.

Galliard, T., 1972, A simple biochemical method for the preparation of high specific activity (^{14}C) acyl-labelled phosphatidylcholine and other phospholipids, *Biochim. Biophys. Acta* **260**:541.

Gammack, D. B., Perrin, J. H., and Saunders, L., 1964, The dispersion of cerebral lipids in aqueous media by ultrasonic irradiation, *Biochim. Biophys. Acta* **84**:576.

Glauert, A. M., and Lucy, J. A., 1969, Electron microscopy of lipids: Effects of pH and fixatives on the appearance of a macromolecular assembly of lipid micelles in negatively stained preparations, *J. Microscopy* **89**:1.

Gobley, M., 1850, Recherches chimiques sur les oeufs de carpe, *J. Pharm. (Paris) Ser. 3* **17**:401.

Goldstein, D. A., and Soloman, A. K., 1960, Determination of equivalent pore radius for human red cells by osmotic pressure measurement, *J. Gen. Physiol.* **44**:11.

Gregoriadis, G., Leathwood, P. D., and Ryman, B. E., 1971, Enzyme entrapment in liposomes, *FEBS Letters* **14**:95.

Hammes, G. G., and Schullery, S. E., 1970, Structure of macromolecular aggregates. 2. Construction of model membranes from phospholipids and polypeptides, *Biochemistry* **9**:2555.

Hammes, G. G., and Tallman, D. E., 1970, Application of the temperature-jump technique to the study of phospholipid dispersions, *J. Am. Chem. Soc.* **92**:6042.

Hanahan, D. J., 1960, "Lipide Chemistry," Wiley, New York.

Hanahan, D. J., Turner, M. B., and Jayko, M. E., 1951, The isolation of egg phosphatidylcholine by an adsorption column technique, *J. Biol. Chem.* **192**:623.

Hanahan, D. J., Dittmer, J. C., and Warashina, E., 1957, A column chromatographic separation of classes of phospholipids, *J. Biol. Chem.* **228**:685.

Hauser, H., 1971, The effect of ultrasonic irradiation of the chemical structure of egg lecithin, *Biochem. Biophys. Res. Commun.* **45**:1049.

Haydon, D. A., 1964, The electrical double layer and electrokinetic phenomena, *in* "Recent Progress in Surface Science," Vol. 1 (J. F. Danielli, K. G. A. Pankhurst, and A. C. Riddiford, eds.), pp. 94–158, Academic Press, New York.

Heap, R. B., Symons, A. M., and Watkins, J. C., 1970, Steroids and their interactions with phospholipids: Solubility, distribution coefficient and effect on potassium permeability of liposomes, *Biochim. Biophys. Acta* **218**:482.

Heap, R. B., Symons, A. M., and Watkins, J. C., 1971, An interaction between oestradiol and progesterone in aqueous solution and in a model membrane system, *Biochim. Biophys. Acta* **233**:307.

Henderson, P. J. F., McGiven, J. D., and Chappell, J. B., 1969, The action of certain antibiotics on mitochondrial erythrocyte and artificial lipid membranes, *Biochem. J.* **111**:521.

Hill, M. W., and Cohen, B. E.,, 1972. A simple method of determining relative permeabilities of liposomes to non-electrolytes. *Biochim. Biophys. Acta* **290**: 403.

Hill, M. W., and Lester, R., 1972, Mixtures of gangliosides and phosphatidylcholine in aqueous dispersions, *Biochim. Biophys. Acta* **282**: 18.

Hillier, A. P., 1970, The binding of thyroid hormones to phospholipid membranes, *J. Physiol.* **211**:585.

Huang, C., 1969, Studies of phosphatidylcholine vesicles. Formation and physical characteristics, *Biochemistry* **8**: 344.

Hubbell, W. L., and McConnell, H. M., 1971, Molecular motion in spin-labelled phospholipids and membranes, *J. Am. Chem. Soc.* **93**:314.

Hübscher, G., Hawthorne, J. N., and Kemp, P., 1960, The analysis of tissue phospholipids: Hydrolysis procedure and results with pig liver, *J. Lipid Res.* **1**:433.

Jacobs, M. H., 1933, The simultaneous measurement of cell permeability to water and to dissolved substances, *J. Cell. Comp. Physiol.* **2**:427.

Johnson, A. R., and Davenport, J. B., eds., 1971, "Biochemistry and Methodology of Lipids," Wiley, New York.

Johnson, J. A., and Wilson, T. A., 1967, Osmotic volume changes induces by a permeable solute, *J. Theoret. Biol.* **17**:304.

Johnson, J. Y., 1932, British Patent No. 417,715, Accepted Oct. 1, 1934, Patent on behalf of I. G. Farbenindustrie Aktiengesellschaft.

Johnson, S. M., and Bangham, A. D., 1969a, Potassium permeability of single compartment liposomes with and without valinomycin, *Biochim. Biophys. Acta* **193**(1):82.

Johnson, S. M., and Bangham, A. D., 1969b, The action of anaesthetics on phospholipid membranes, *Biochim. Biophys. Acta* **193**(1):92.

Johnson, S. M., and Buttress, N., 1972, The osmotic insensitivity of sonicated liposomes and the density of phospholipid–cholesterol mixtures. *Biochim. Biophys. Acta* **307**: 20.

Johnson, S. M., and Miller, K. W., 1970, Antagonism of pressure and anaesthesia, *Nature* **228**:75.

Johnson, S. M., Bangham, A. D., Hill, M. W., and Korn, E. D., 1971, Single bilayer liposomes, *Biochim. Biophys. Acta* **233**:820.

Jost, P., Libertini, L. J., Hebert, V. C., and Griffith, O. H., 1971, Lipid spin labels in lecithin multilayers. A study of motion along fatty acid chains, *J. Mol. Biol.* **59**:77.

Junger, E., and Reinauer, H., 1969, Liquid crystalline phases of hydrated phosphatidylethanolamine, *Biochim. Biophys. Acta* **183**:304.

Junger, E., Hahn, M. H., and Reinauer, H., 1970, The structure of lysolecithin–water phases. Negative staining and optical diffraction analysis of the electron micrographs, *Biochim. Biophys. Acta* **211**:381.

Kaufman, S., Steim, J. M., and Gibbs, J. H., 1970, Nuclear relaxation in phospholipids and biological membranes, *Nature* **225**:743.

Kavanau, J. L., 1965, "Structure and Function in Biological Membranes," Vol. I and II, Holden-Day, San Francisco.

Kerker, M., 1969, "The Scattering of Light and Other Electromagnetic Radiation," Academic Press, New York.

Kimelberg, H. K., and Papahadjopoulos, D., 1971a, Phospholipid–protein interactions: Membrane permeability correlated with monolayer "penetration," *Biochim. Biophys. Acta* **233**:805.

Kimleberg, H. K., and Papahadjopoulos, D., 1971b, Interactions of basic proteins with phospholipid membranes, *J. Biol. Chem.* **246**:1142.

Kinsky, S. C., 1970, Antibiotic interaction with model membranes, *Ann. Rev. Pharmacol.* **10**:119.

Kinsky, S. C., 1972, Antibody–complement interaction with lipid model membranes, *Biochim. Biophys. Acta* **265**:1.

Kinsky, S. C., Haxby, J., Kinsky, C. B., Hemel, R. A., and van Deenen, L. L. M., 1968, Effect of cholesterol incorporation in sensitivity of liposomes in the polyene antibiotic, Filipin, *Biochim. Biophys. Acta* **152**:174.

Klein, R. A., 1970*a*, The detection of oxidation in liposome preparations, *Biochim. Biophys. Acta* **210**:486.

Klein, R. A., 1970*b*, The large scale preparation of unsaturated phosphatidylcholines from egg yolk, *Biochim. Biophys. Acta* **219**:496.

Klein, R. A., 1972, Mass spectrometry of the phosphatidyl amino alcohols: detection of molecular species and use of low voltage spectra and metastable scanning in the elucidation of structure, *J. Lipid Res.* **13**:672.

Klein, R. A., Moore, M. J., and Smith, M. W., 1971, Selective diffusion of neutral amino acids across lipid bilayers, *Biochim. Biophys. Acta* **233**:420.

Kornberg, R. D., and McConnell, H. M., 1971, Inside–outside transitions of phospholipids in vesicle membranes, *Biochemistry* **10**:1111.

Kornberg, R. D., McNamee, M. G., and McConnell, H. M., 1972, Measurement of transmembrane potentials in phospholipid vesicles, *Proc. Natl. Acad. Sci. (U.S.)* **69**: 1508.

Kruyt, H. R., 1948, "Colloid Science," Vol. 2, Chap. IX, Elsevier, Amsterdam.

Ladbrooke, B. D., and Chapman, D., 1969, Thermal analysis of lipids, proteins and biological membranes. A review and summary of some recent studies, *Chem. Phys. Lipids* **3**:304.

Lawrence, A. S. C., 1969, Lyotropic mesomorphisms in lipid–water systems, *Mol. Cryst. Liquid Cryst.* **7**:1.

Lea, C. H., Rhodes, D. N., and Stoll, R. D., 1955, Phospholipids. 3. On the chromatographic separation of glycerophospholipids, *Biochem. J.* **60**:353.

Lee, A. G., Birdsall, N. J. M., Levine, Y. K., and Metcalfe, J. C., 1972, High resolution proton relaxation studies of lecithins, *Biochim. Biophys. Acta* **255**:43.

Lees, M., Folch, J., Sloane-Stanley, G. H., and Carr, S., 1959, A simple procedure for the preparation of brain sulphatides, *J. Neurochem.* **4**:9.

Lehninger, A. L., Carafoli, E., and Rossi, C. S., 1967, Ion movements in mitochondrial systems, *Advan. Enzymol.* **29**:259.

Lester, R., Hill, M. W., and Bangham, A. D., 1972, Molecular mechanism of Tay-Sachs disease, *Nature* **236**:32.

Levine, Y. K., and Wilkins, M. H. F., 1971, Structure of oriented lipid bilayers, *Nature New Biol.* **230**:69.

Luzzati, V., and Husson, F., 1962, The structure of the liquid-crystalline phases of lipid–water systems, *J. Cell Biol.* **12**:207.

MacDonald, R. C., and Bangham, A. D., 1972, Comparison of double layer potentials in lipid monolayers and lipid bilayer membranes, *J. Membrane Biol.* **7**:29.

Mangold, H. K., 1969, *in* "Thin Layer Chromatography. A Laboratory Handbook" (E. Stahl, ed.), Springer-Verlag, Berlin.

Marinetti, G. V., 1967, "Lipid Chromatographic Analysis," Vol. 1, M. Dekker, New York.

McConnell, H. M., and McFarland, B. G., 1970, Physics and chemistry of spin labels, *Quart. Rev. Biophys.* **3**:91.

McElhaney, R. N., de Gier, J., and Van Deenen, L. L. M., 1970, The effect of alterations in fatty acid composition and cholesterol content on the permeability of *Mycoplasma laidlawii* B cells and derived liposomes, *Biochim. Biophys. Acta* **219**:245.

Melchior, D. L., Morowitz, H. J., Sturtevant, J. M., and Tsong, T. Y., 1970, Characterization of the plasma membrane of *Mycoplasma laidlawii*, *Biochim. Biophys. Acta* **219**:114.

Metcalfe, J. C., Birdsall, N. J. M., Feeney, J., Lee, A. G., Levine, Y. K., and Partington, P., 1971, ^{13}C NMR spectra of lecithin vesicles and erythrocyte membranes, *Nature* **233**:199.

Miyamoto, V. K., and Stoeckenius, W., 1971, Preparation and characteristics of lipid vesicles, *J. Membrane Biol.* **4(3)**:252.

Mueller, P., Rudin, D. O., Tien, H. T., and Wescott, W. C., 1962a, Reconstitution of excitable membrane structure *in vitro*, *Circulation* **26**:1167.

Mueller, P., Rudin, D. O., Tien, H. T., and Wescott, W. C., 1962b, Reconstitution of cell membrane structure *in vitro* and its transformation into an excitable system, *Nature* **194**:949.

O'Neill, M. J., 1964, The analysis of a temperature controlled scanning calorimeter, *Anal. Chem.* **36**:1238.

Oparin, A. I., 1924, "Proiskhozhdenie Zhizni," Moscow; 1938, "The Origin of Life on Earth," Macmillan, New York; 1953, "The Origin of Life," Dover, New York; 1957, "The Origin of Life on Earth," Academic Press, New York.

Owen, J. D., Hemmes, P., and Eyring, E. M., 1970, Light scattering temperature jump relaxations in mixed solvent suspensions of phosphatidylcholine vesicles, *Biochim. Biophys. Acta* **219**:276.

Palmer, K. J., and Schmitt, F. O., 1941, X-ray diffraction studies of lipide emulsions, *J. Cell. Comp. Physiol.* **17**:385.

Pangborn, M. C., 1951, A simplified purification of lecithin, *J. Biol. Chem.* **188**:471.

Papahadjopoulos, D., 1971, Na$^+$–K$^+$ discrimination by "pure" phospholipid membranes, *Biochim. Biophys. Acta* **241**:254.

Papahadjopoulos, D., 1972, Studies on the mechanism of action of local anaesthetics with phospholipid model membranes, *Biochim. Biophys. Acta* **265**:69.

Papahadjopoulos, D., and Bangham, A. D., 1966, Permeability of phosphatidylserine liquid crystals to univalent ions, *Biochim. Biophys. Acta* **126**:185.

Papahadjopoulos, D., and Miller, N., 1967, Phospholipid model membranes. I. Structural characteristics of hydrated liquid crystals, *Biochim. Biophys. Acta* **135**:624.

Papahadjopoulos, D., and Watkins, J. C., 1967, Phospholipid model membranes. II. Permeability properties of hydrated liquid crystals, *Biochim. Biophys. Acta* **135**:639.

Papahadjopoulos, D., Yin, E. T., and Hanahan, D. J., 1964, Purification and properties of bovine factor X: Molecular changes during activation, *Biochemistry* **3**:1931.

Parker, F., and Peterson, N. F., 1965, Quantitative analysis of phospholipids and phospholipid fatty acids from silica gel thin-layer chromatograms, *J. Lipid Res.* **6**: 455.

Penick, R. J., Meisler, M. H., and McCluer, R. H., 1966, Thin-layer chromatographic studies of human brain gangliosides, *Biochim. Biophys. Acta* **116**:279.

Penkett, S. A., Flook, A. G., and Chapman, D., 1968, Physical studies of phospholipids. IX. Nuclear resonance studies of lipid–water systems, *Chem. Phys. Lipids* **2**:273.

Quarles, R. H., and Dawson, R. M. C., 1969, The distribution of phospholipase D in developing and mature plants, *Biochem. J.* **112**:787.

Redwood, W. R., Takashima, S., Schwan, H. P., and Thompson, T. E., 1972, Dielectric studies of homogenous phosphatidylcholine vesicles, *Biochim. Biophys. Acta* **255**:557.

Reeves, J. P., and Dowben, R. M., 1969, Formation and properties of thin-walled phospholipid vesicles, *J. Cell. Physiol.* **73**:49.

Reeves, J. P., and Dowben, R. M., 1970, Water permeability of phospholipid vesicles, *J. Membrane Biol.* **3**:123.

Reinert, J. C., and Steim, J. M., 1970, Calorimetric detection of a membrane–lipid phase transition in living cells, *Science (N.Y.)* **168**:1580.

Reman, F. C., Demel, R. A., de Gier, J., van Deenen, L. L. M., Eibl, H., and Westphal, O., 1969, Studies on the lysis of red cells and bimolecular lipid leaflets by synthetic lysolecithins, lecithins and structural analogs, *Chem. Phys. Lipids* **3**:221.

Rendi, R., 1965, Osmotic properties of phospholipid suspensions, *J. Cell Biol.* **27**:83A.

Rendi, R., 1967, Water extrusion in isolated subcellular fractions. VI. Osmotic properties of swollen phospholipid suspensions, *Biochim. Biophys. Acta* **135**:333.

Rhodes, D. N., and Lea, C. H., 1956, in "Biochemical Problems of Lipids," pp. 77–79, Butterworths, London.

Rhodes, D. N., and Lea, C. H., 1957, Phospholipids. 4. On the composition of hen's egg phospholipids, *Biochem. J.* **65**:526.

Robinson, N., 1960, A light scattering study of lecithin, *Trans. Faraday. Soc.* **56**:1260.

Rouser, G., Bauman, A. J., Kritchevsky, G., Heller, P., and O'Brien, J. S., 1961, Quantative chromatographic fractionation of complex lipid mixtures: Brain lipids, *J. Am. Oil Chem. Soc.* **38**:544.

Rouser, G., Kritchevsky, G., Heller, D., and Lieber, E., 1963, Lipid composition of beef brain, beef liver, and the sea anemone: Two approaches to quantitive fractionation of complex lipid mixtures, *J. Am. Oil Chem. Soc.* **40**:425.

Saha, J., Papahadjopoulos, D., and Wenner, C. E., 1970, Studies on model membranes. I. Effects of Ca^{2+} and antibiotics on permeability of cardiolipin liquid-crystalline vesicles, *Biochim. Biophys. Acta* **196**:10.

Salsbury, N. J., and Chapman, D., 1968, Physical studies of phospholipids. VIII. Nuclear magnetic resonance studies of diacyl-l-phosphatidylcholines (lecithins), *Biochim. Biophys. Acta* **163**:314.

Saunders, L., 1963, Lecithin–cholesterol sols, *J. Pharm. Pharmacol.* **15**:155.

Saunders, L., Perrin, J., and Gammack, D. B., 1962, Ultrasonic irradiation of some phospholipid sols, *J. Pharm. Pharmacol.* **14**:567.

Scarpa, A., and de Gier, J., 1971, Cation permeability of liposomes as a function of the chemical composition of the lipid bilayers, *Biochim. Biophys. Acta* **241**:789.

Schmitt, F. O., Bear, R. S., and Clark, G. L., 1935, X-ray studies of nerve, *Radiology* **25**:131.

Schulman, J. H., and Rideal, E. K., 1931, Surface potentials with alpha particle sources, *Proc. Roy. Soc.* **A130**:259.

Selwyn, M. J., Dawson, A. P., Stockdale, M., and Gains, N., 1970, Chloride–hydroxide exchange across mitochondrial, ethryocyte and artificial membranes mediated by trialkyl and triphenyltin compounds, *Europ. J. Biochem.* **14**:120.

Sessa, G., and Weissmann, G., 1968, Phospholipid spherules (liposomes) as a model for biological membranes, *J. Lipid Res.* **9**:310.

Sessa, G., and Weissmann, G., 1970, Incorporation of lysozyme into liposomes, *J. Biol. Chem.* **245**:3295.

Sessa, G., Freer, J. H., Colacicco, G., and Weissmann, G., 1969, Interaction of a lytic poly-peptide melittin with lipid membrane systems, *J. Biol. Chem.* **244**:3575.

68   A. D. Bangham, M. W. Hill, and N. G. A. Miller

Seufert, W. D., 1970, Model membranes: Spherical shells bounded by one bimolecular layer of phospholipids, *Biophysik* **7**:60.

Sha'afi, R. I., Rich, G. T., Mikulecky, D. C., and Solomon, A. K., 1970, Determination of urea permeability in red cells by minimum method, *J. Gen. Physiol.* **55**:427.

Sigler, K., and Janáček, K., 1971, The effect of non-electrolyte osmolarity on frog oocytes. 1. Volume changes, *Biochim. Biophys. Acta* **241**:528.

Singer, M. A., and Bangham, A. D., 1971, The consequences of inducing salt permeability in liposomes, *Biochim. Biophys. Acta* **241**:687.

Singleton, W. S., Gray, M. S., Brown, M. L., and White, J. L., 1965, Chromatographically homogeneous lecithin from egg phospholipids, *J. Am. Oil Chem. Soc.* **42**:53.

Small, D. M., 1967, Phase equilibria and structure of dry and hydrated egg lecithin, *J. Lipid Res.* **8**:551.

Small, D. M., 1970, Surface and bulk interactions of lipids and water with a classification of biologically active lipids based on their interactions, *Fed. Proc.* **29**:1320.

Strecker, A., 1868, Über das Lecithin, *Ann. Chem. Pharm.* **148**:77.

Sweet, C., and Zull, J. E., 1969, Activation of glucose diffusion from egg lecithin liquid crystals by serum albumin, *Biochim. Biophys. Acta* **173**:94.

Tedeschi, H., and Harris, D. L., 1955, The osmotic behaviour and permeability to non-electrolytes of mitochondria, *Arch. Biochem. Biophys.* **58**:52.

Thannhauser, S. J., and Setz, P., 1936, Studies on animal lipids. XI. The Reineckate of the polydiaminophosphatide from spleen, *J. Biol. Chem.* **116**:527.

Thudichum, J. L. W., 1874, Researches in the chemical constitution of the brain, *in* Rep. Med. Off. Priv. Council, New Ser. No. 3, App. 5, p. 133, Eyre and Spottiswoode, London.

Tinker, D. O., 1972, Light scattering by phospholipid dispersions: Theory of light scattering by hollow spherical particles, *Chem. Phys. Lipids* **8**:230.

Tinker, D. O., and Pinteric, L., 1971, On the identification of lamellar and hexagonal phases in negatively stained phospholipid–water systems, *Biochemistry* **10**:860.

Trosper, T., and Raveed, D., 1970, Chlorophyll-a-containing liposomes, *Biochim. Biophys. Acta* **223**:463.

Watson, E. S., O'Neill, M. J., Justin, J., and Brenner, N., 1964, A differential scanning calorimeter for quantitative differential thermal analysis, *Anal. Chem.* **36**:1233.

Weissmann, G., 1971, The molecular basis of gout, *Hosp. Practice*, July, 43.

Weissmann, G., and Rita, G. A., 1972, The molecular basis of gouty inflammation: Interaction of monosodium urate crystals with lysosomes and liposomes. *Nature (Lond.)*, **240**:167.

Weissmann, G., and Sessa, G., 1967, The action of polyene antibiotic on phospholipid cholesterol structure, *J. Biol. Chem.* **242**:616.

Wheeler, K. P., and Whittam, R., 1970. The involvement of phosphatidylserine in adenosine triphosphatase activity of the sodium pump, *J. Physiol. (Lond.)* **207**:303.

Wilkins, M. H. F., Blaurock, A. E., and Engelman, D. M., 1971, Bilayer structure in membranes, *Nature New Biol.* **230**:72.

Chapter 2

Thermodynamics and Experimental Methods for Equilibrium Studies with Lipid Monolayers

N. L. GERSHFELD

National Institute of Arthritis, Metabolism, and Digestive Diseases
National Institutes of Health
Bethesda, Maryland

1. INTRODUCTION

Lipid monolayers on water have long been favored as a structural model of the cell membrane. However, despite intensive efforts, the insights which the lipid films have contributed to the understanding of membrane structure and function have been largely qualitative. Thus the importance of molecular shape and orientation at water surfaces (Hardy, 1912; Langmuir, 1917) and the existence of monolayer states which are analogous to those states which exist in bulk (Devaux, 1913) were established with studies of single-component lipid films (Adam and Jessop, 1926). But questions pertaining to the energetics of the lipid–lipid and polar group interactions and the contribution of water to film structure were left largely speculative.

   Of more immediate interest are the studies with surface film mixtures of cholesterol and phospholipids to explain the role of cholesterol in cell membranes. Attempts to characterize the interactions in these two-component films from the apparent changes in molecular packing have also resulted in a very qualitative picture of the energetics of mixing. Moreover, recent studies have raised serious questions about the validity of using changes in packing area for describing the interactions in lipid monolayer mixtures (Gershfeld and Pagano, 1972*a*).

In addition to the energetics of film structure, film penetration has been of active interest to workers in the membrane field (Schulman and Hughes, 1935; Gaines, 1966). Typically, a water-soluble, surface-active component ("penetrant") is introduced into the solution beneath the insoluble lipid monolayer system, and the resulting change in monolayer properties is interpreted in terms of the "penetrant"–film interaction. The film penetration studies of relevance to the membrane system utilized pharmacological agents as the penetrant, presumably to simulate conditions for testing drug–membrane receptor interactions. The majority of these studies have been marked by inconclusive results.

The usefulness of any model system is obviously limited by the extent to which the system is defined. For each of the systems cited, inconclusive results which have been obtained can be related directly to the absence of a well-defined system. In this chapter, an analysis of the one-component, two-component, and "penetrated" monolayer systems is presented, which will provide a theoretical basis for the experimental study of these systems. The analysis will include a discussion of some of the conceptual difficulties to be encountered when studying lipid monolayers. The chapter is divided into two sections: the first contains an analysis of the thermodynamics of lipid monolayers for evaluating those parameters which will define the monolayer system; the second deals with the experimental details for the measurement of the essential parameters treated in the first section.

For the historical background and summaries of the earlier studies, both Rideal's (1930) and Adam's (1941) books provide well-written and useful descriptions of the work of that period. More recent material has been presented by Gaines (1966). For rigorous treatment of the thermodynamics of surfaces, the reader is referred to Defay et al. (1966). Finally, some of the conclusions to be summarized in this chapter have been treated more completely in a recent series of publications (Gershfeld, 1972; Gershfeld and Pagano, 1972a,b; Pagano and Gershfeld, 1972a,b).

2. THERMODYNAMICS

2.1. The Monolayer System

According to the Gibbs adsorption isotherm (Gibbs, 1948), the composition of an interface is a function of the chemical potentials of the components in the equilibrium bulk phase. Thus,

$$- d\gamma = \Sigma \Gamma_i \, d\mu_i \tag{1}$$

where γ is the surface tension of the solution, Γ_i is the surface excess* concentration of component i, and μ_i is the chemical potential.

Another form of the Gibbs equation is in terms of the surface pressure $\Pi = \gamma_0 - \gamma$, where γ_0 is the surface tension of the solution in the absence of surface-active components:

$$d\Pi = \Sigma \Gamma_i \, d\mu_i \tag{2}$$

In the case of the air–water system, the lipid which is surface active may appear in each bulk phase—the subsolution and the vapor—as well as in the surface. Hence, when the monolayer is studied systematically, it must be viewed within the general framework of the entire system—the monolayer, subsolution, and vapor.

A common method of preparing a monolayer is to dissolve the lipid directly in the aqueous phase; adsorption of the surface-active lipid to the surface is always in accordance with the Gibbs adsorption isotherm. In the event the lipid is virtually insoluble in the water, a second method has been used—the lipid is first dissolved in a volatile, water-insoluble solvent, and the solution is then spread directly on the water surface. After the solvent has evaporated, only the lipid remains on the surface. The latter method—the spread film—has been favored in many studies because most of the membrane lipids are extremely insoluble in water and readily form spread films.

A prevalent misconception is that the spread films are insoluble. Actually, many spread films dissolve in the subsolution, and in some instances evaporation of the lipid from the aqueous phase to the vapor also has been observed (Mansfield, 1959; Roylance and Jones, 1960; Brooks and Alexander, 1960; Muramatsu and Ono, 1971). Until recently, the slight solubility of these spread films has been regarded as an experimental

---

* The surface excess concentration refers to the amount of adsorbed component which is in excess of the amount that would be present at the interface if the concentration in the bulk were uniformly maintained into the region of the surface. Because of surface forces, the amount in the surface may be greater or less than that which would appear in the absence of surface force; thus it is possible for Γ_i to be positive, zero, or negative. In the case of spread films which are extremely insoluble in the subsolution, Γ is the amount spread and is usually treated as the surface concentration without the term "excess" included. Since most of our discussion here will concern spread films, the term "surface concentration" will be used exclusively. In those instances where soluble surface-active compounds are employed, the excess property will be clearly indicated.

difficulty, sometimes referred to as film "instability." In reality, the desorption phenomenon is in accordance with predictions of the Gibbs equation (equation 1) and has been utilized to determine the chemical potential of the lipid in the surface (Gershfeld and Patlak, 1966; Patlak and Gershfeld, 1967; Gershfeld, 1968).

Even for the most stable of spread films, a certain amount of film loss to the adjacent bulk phase is to be expected, in principle. To visualize how this desorption will influence the measured properties of the surface film, it is instructive to consider the mechanism of film desorption (Ter Minassian-Saraga, 1955). It has been shown that immediately after spreading, desorption of some of the spread film occurs into a very thin region of the subsolution. The amount of material which desorbs immediately is determined by the Gibbs equation. This is followed by diffusion into the bulk region of the subsolution. The latter process is generally the rate-determining step and will depend on the concentration of lipid in the thin region beneath the film. For very slightly soluble lipids, this process will be extremely slow, and within the normal range of experimental times loss of film by diffusion will be very small. The net affect is that the apparently insoluble lipid spread film is in equilibrium with only a very thin region of the subsolution. Clearly, for the more soluble lipids desorption occurs at a more rapid rate. Thus there is a spectrum of behavior for the lipids depending on the relative solubility of the lipid in the aqueous subsolution.

In principle, films formed either by adsorption or by spreading should be identical. To demonstrate that the films are identical, the following is taken from an experiment published recently (Gershfeld and Pagano, 1972b).

Cholesterol is generally considered to be extremely insoluble in water, and it also readily forms spread films. When a saturated solution of cholesterol in water is prepared, the measured surface tension is 33 dyn/cm; if the surface concentration is measured independently by using radiotracers of cholesterol (see Section 3.3), the experimental value is 4.5×10^{-10} mole/cm^2. If the spread film of cholesterol is compressed to a surface concentration of 4.5×10^{-10} mole/cm^2 (equivalent to about 37 Å^2/molecule), the surface tension measured will be 33 ± 1 dyn/cm, in agreement with the values obtained for the adsorbed film. Thus the two methods yield identical films, even in the case of an extremely insoluble compound such as cholesterol.

The surface pressure ($\Pi = \gamma_0 - \gamma$) of the system of film in equilibrium with excess (saturated) lipid is known as the "equilibrium spreading pressure" (Π_e), since it may also be obtained by placing a crystal of the lipid directly on the water surface. The lipid will spread spontaneously over the surface as well as dissolve in the subsolution. The former process is orders of magnitude faster than the latter. Therefore, a convenient method for the preparation of a saturated surface is to place a crystal of lipid on the surface. Since the surface is saturated with respect to the lipid, at equilibrium the bulk solution must therefore also be saturated.

In summary, for the monolayer system all phases must be considered in equilibrium with the monolayer. In the case of spread films, the film may be considered as being insoluble under the conditions where no desorption or surface evaporation can be detected for the duration of the experiment. If film instability is evident, it is most probably due to the drift toward the equilibrium state described by the Gibbs adsorption isotherm.

The scope of the monolayer system having been defined, it is now appropriate to deal with what is probably the most confused and experimentally difficult aspect of the monolayer system, namely, the related problems of equilibrium and film homogeneity. The question is of extreme importance, for the very essence of the monolayer as a general model system depends on establishing whether the system is in an equilibrium state. Thermodynamic parameters obtained under equilibrium conditions are state functions and are independent of the processes utilized to reach that state. However, if the system is dependent on the pathway selected to reach that state, it is obvious that the generality of the parameters describing that state will be severely restricted.

2.2. Equilibrium and Supercompressed States

With bulk systems, the establishment of equilibrium is generally accomplished by varying the temperature, pressure, and composition around the equilibrium values of these parameters, thereby approaching the equilibrium state by different experimental pathways. Moreover, the system may be stirred, and it may be isolated and left indefinitely so that sufficient time is allowed to reach equilibrium. In addition, the gravitational field will facilitate the separation of phases, which can usually be detected easily.

For lipid films, the question of establishing equilibrium is complicated by experimental difficulties. For example, it is rare that the air–water system can be left for prolonged periods without serious contaminations appearing at the interface. There is also a natural reluctance to stir the surface for fear of disrupting the system irreversibly. If phase separation in the surface occurs, it is virtually impossible to identify the separate phases in the surface by visual means. Coupled with these difficulties is the fact that for many of the lipid films of interest the surface viscosity is extremely high (described picturesquely as being equivalent to that for butter; Davies and Rideal, 1961) and is likely to contribute time-dependent surface phenomena and retard the normal drift of the system toward its equilibrium state.

To illustrate some of the problems associated with establishing equilibrium conditions in monolayers, consider the following two systems which are representative of many monolayer studies in the literature. The first is the one-component monolayer on water, stearic acid, and the second is the binary mixture of cholesterol and dipalmitoyl-1-α-lecithin (DPL) on water.

2.2.1. One Component Systems

Spread films of lipids such as stearic acid have been studied typically by compressing the surface film from some large area, where the surface pressure (Π) is close to zero, to an area where the film appears to collapse, as indicated by rapid and marked changes in the surface pressure. The latter is known as the "collapse pressure," (Π_c), and can be as high as 50–60 dyn/cm. A typical isotherm for stearic acid is shown in Fig. 1; the region from 25 to 19 $\mathring{A}^2$/molecule is known as the "condensed region." The state of collapse is considered to be the point at which the monolayer has crumbled to form excess bulk (crystalline) stearic acid where the presence of crystals can be verified by ultramicroscopy (Harkins, 1945). In principle, the value of Π when crystals first appear should be the equilibrium spreading pressure (Π_e) for stearic acid, the value one would obtain if crystals of stearic acid were in equilibrium with monolayer on water. For stearic acid, Π_e is about 7 dyn/cm (Heikkila et al., 1970), whereas the collapse pressure is considerably higher (see Fig. 1). Moreover, the very high values for Π_c can only be reached by compressing a spread film, and by no other process, e.g., spontaneous spreading from the crystal.* It is therefore apparent that the region of the isotherm which exceeds the equilibrium spreading pressure describes a nonequilibrium state, a state which has characteristics of metastability.

Demonstration of the existence of metastable states for films such as

---

* Reports of collapsed films which are allowed to respread in the absence of stable bulk phase yielding surface pressures which exceed Π_e (Brooks and Alexander, 1962a, b; Heikkila et al., 1970) can be attributed to a process which is analogous to supersaturation, where the presence of extremely small crystals leads to an increase in the solubility of the material. This is also a metastable condition where larger particles are formed at the expense of the small ones and the solubility ultimately approaches the equilibrium state.

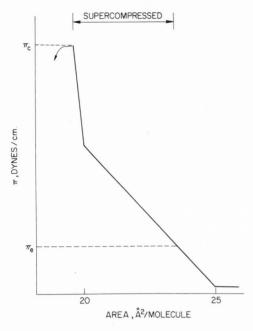

Fig. 1. Surface pressure–area isotherm for the condensed region of a
stearic acid monolayer. The relative
positions of the equilibrium spreading pressure (Π_e) and the collapse
pressure (Π_c) are indicated. When
the surface pressure (Π) exceeds
Π_e, the film becomes metastable;
this state has been called "supercompressed."

stearic acid may be gleaned from data presented by Rabinovitch *et al.* (1960),
who have shown that if films of stearic acid which have been compressed
beyond Π_e are allowed to stand at constant film area, the surface pressure
falls and approaches the equilibrium spreading pressure. This decrease in
surface pressure is not due to dissolution or film evaporation, because when
the film area is increased to some large value and the film is allowed to stand
for as much as 24 hr and then compressed again, the original isotherm is
obtained. This can only mean that the original decrease in surface pressure
is due to formation of a bulk state (crystalline), which if allowed to re-
equilibrate at the larger film areas reforms the monolayer.

It is clear that for this example the isotherm region beyond Π_e is that
of an unstable nonequilibrium state. Since these states can only be produced
by film compression, they will be called "supercompressed states" by analogy
with the supercooled states observed with bulk liquids.

The supercompressed states undoubtedly owe their apparent stability
to the fact that condensed films are quite viscous. The existence of supercompressed states in many reports may be inferred from the need to use
reproducible rates of film compression in order to obtain reproducible
isotherms. In general, it is best to avoid exceeding Π_e, unless one is interested
in studying metastable systems.

2.2.2. Two-Component System

The second system for analysis is the binary lipid mixture as a spread monolayer. The problem of establishing that the system is at equilibrium is now compounded by the presence of a second lipid component in the surface film. Thus in addition to the problem of supercompression there are the complications inherent in the mixing of two components, e.g., immiscibility and phase separation. For these mixed films, most of the studies have been restricted to the condensed region of the isotherm (Fig. 1); Crisp (1949) has analyzed possible phase relations for these systems.

To illustrate the difficulties inherent in the study of condensed multi-component lipid monolayers, a mixing experiment was performed by two different approaches. In one, cholesterol and DPL at a fixed molar ratio were spread on the water surface from a common volatile solvent. The Π–A isotherm indicates that Π is a continuous function of the film area (van Deenen *et al.*, 1962), which indicates that the spread film is homogeneous (see Section 2.3). Thus from the spread film experiment it may be concluded that the two lipid components are miscible at all film pressures.

The second approach (Gershfeld and Pagano, 1972*b*) for forming mixed lipid films is to prepare saturated solutions of the two lipid components in water. Excess crystalline DPL and cholesterol crystals were added to water, and the mixture was stirred vigorously; the surface tension and surface composition were monitored periodically until no change was detected even after additional lipid material was added with stirring. According to the phase rule of Gibbs (to be discussed in detail in the following section), the surface film can have only one surface phase. Hence if the equilibrium composition of the surface is measured, the absence of either component would signify that the two components are immiscible in the surface. Cholesterol in the surface was measured by radiotracers (see Section 3.3), and the equilibrium surface concentration was found to be identical with that for a saturated solution of pure cholesterol. In other words, the second lipid component DPL was completely excluded from the surface, indicating complete immiscibility of cholesterol and DPL in surface films. The result of this experiment obviously contradicts that for the spread film experiment, which had indicated complete miscibility of these components at all surface pressures. Since the second experiment was performed under conditions which are likely to reach equilibrium (stirring), and in view of the possible occurrence of supercompression in spread films and the high surface viscosity in condensed films, it must be assumed that the results of the spread film experiment are for a nonequilibrium system. It should be noted that if excess cholesterol is added to the spread mixed film and the system is then stirred, the resulting film contains only cholesterol. The reason for the presence of cholesterol and not DPL in these experiments can be derived from thermodynamics; in essence, the compound with the highest equilibrium spreading pressure will preferentially adsorb in the surface when the components are immiscible (Gershfeld and Pagano, 1972*b*).

Mixed films of cholesterol and phospholipids are usually prepared by spreading from a common solvent. Attempts to use the average area per film molecule as a measure of the interactions which are presumed to occur in the spread mixed films must be questioned on the basis that the calculations assume that the films are homogeneous. In view of the possibility of phase separation and nonequilibrium conditions in the spread films, especially in regions of high compression, the significance of the changes in packing areas of the spread mixed films must remain open.

For single-component spread films, the question of film heterogeneity will always be present if the equilibrium spreading pressure is exceeded, and as a minimum requirement Π_e should not be exceeded if equilibrium studies are contemplated. For binary lipid mixtures, the problem is more complex, since the possibility exists of immiscibility of the surface components. Since there is no theoretical method for predicting the conditions of miscibility and the experimental surface techniques cannot detect phase separation reliably, we must rely on the phase rule of Gibbs to limit the study of monolayers to situations where the phase relations are unequivocal. A brief discussion of the phase rule as it applies to monolayers will now be presented.

2.3. The Gibbs Phase Rule and Phase Separation in Monolayers

The difficulty of establishing visually the phase relations in binary lipid monolayers can be partially obviated by restricting the experimental conditions to those situations where only a minimum number of possibilities can occur. The phase rule of Gibbs relates the number of experimental variables with the number of phases in the system at equilibrium; the explicit relation for monolayers (Defay, 1932; Crisp, 1949) is given by

$$F = C - P^b - P^s + 3 \qquad (3)$$

where C is the number of components of the system, P^b is the number of bulk phases, P^s is the number of surface phases, and F (the degrees of freedom) is the number of intensive variables of the system such as temperature, pressure,

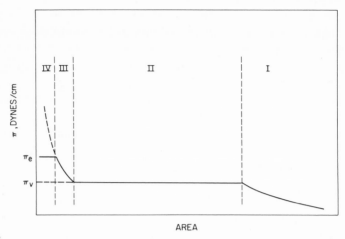

Fig. 2. **Surface pressure–area isotherm for a single-component monolayer.** Region I, gaseous state; II, transition region; III, condensed state; and IV, supercompressed state. Π_v and Π_e are the surface vapor and equilibrium spreading pressures, respectively.

concentration, and surface pressure, which must be fixed in order that the state of the system be completely defined. Of particular interest is the fact that the degrees of freedom in the system unequivocally define the relation existing between the number of components and the number of phases in the system. This relation may be used to define the experimental conditions under which the phase behavior of the monolayers may be established. It is important to emphasize that the phase rule can be applied only to systems which are at equilibrium. Hence the existence of nonstable or metastable states cannot be treated by the phase rule. To illustrate its use, the two examples considered previously, single- and binary-component monolayers, will be examined in detail.

For a single-component lipid film, such as palmitic acid, the isotherm presented in Fig. 1 represents only a small portion of the total Π–A relation describing the behavior of palmitic acid films on water. A complete isotherm, drawn schematically in Fig. 2, will show at least four distinct regions.

2.3.1. Region I

In region I, Π is a function of A (or surface concentration); since temperature and pressure are constant, Π is the only independent intensive variable and hence $F = 1$. From the phase rule, since $C = 2$ (lipid and water) and $P^b = 1$ (water, ignoring air since it represents one component and one bulk phase, thereby cancelling out in the phase rule), P^s must equal 1. Hence region I is a single, homogeneous lipid surface phase. In this region of the isotherm, the film behaves like a gas. In the limit of very large molecular areas $(A = 10^5$ Å^2/molecule), the isotherm may be represented by the relation

$$\Pi A = kT$$

which is formally equivalent to the ideal gas law $pv = RT$ (Adam and Jessop, 1926). Thus region I is termed the "gaseous region" of the isotherm.

2.3.2. Region II

In region II, Π is a constant Π_v and independent of the film area A, and the state of the system is independent of Π. Consequently, $F = 0$, and $P^s = 2$. Thus two surface phases coexist in region II. Region II is bounded by the gaseous region at the high area end and by the condensed region (region III) (see also Fig. 1) at the lower area end; thus the two surface phases in region II are the gaseous and condensed phases. The similarity of this portion of the isotherm to the transition region of the pressure–volume relation for gases in which the vapor undergoes a transformation to the liquid led to the introduction of the term "transition region", where the constant value of the surface pressure is called the "surface vapor pressure", (Π_v) (Adam and Jessop, 1926). Characteristic values for Π_v are given in Table I.

Table I. Surface Vapor Pressure Π_v and A_c Values for Various Lipids[a] ($T = 298\,°$K)

Lipid	Π_v ($\pm$0.001 dyn/cm)	A_c (Å^2/molecule)
Cholesterol	0.006	40
Stearic acid	0.012	24.5
Stearyl alcohol	0.010	22.0
Oleic acid	0.095	58
Oleyl alcohol	0.085	49
Palmitic acid	0.076	25.5
Methyl stearate	0.036	22

[a] Π_v values from Gershfeld and Pagano (1972a). A_c values from Gaines (1966).

2.3.3. Region III

In region III, Π is a function of the film area and, as with region I, the film is comprised of a single surface phase. This is the region represented in the isotherm shown in Fig. 1 and is the lipid in a condensed state. The lipid in this phase has properties strikingly similar to those of bulk liquids and thus has been given various names such as "liquid–condensed" or "liquid–expanded" to signify the essentially liquid-like nature of the monolayers (Harkins, 1945). Approximate values for A_c, the area/molecule at which the surface pressure deviates from Π_v, are given in Table I for various lipids.

It should be noted that various condensed film states have been reported (for reviews, see Harkins, 1952; Gaines, 1966). The exact nature of these states is still disputed, and only a qualitative and descriptive discussion of these states can be given at this time; the interested reader is referred to the literature for further information (Harkins, 1952; Gaines, 1966).

2.3.4. Region IV

In region IV, Π is a constant, the equilibrium spreading pressure Π_e, and $F = 0$. From our previous discussion (Section 2.2), excess lipid phase is present (e.g., crystals of stearic acid), $P^b = 2$ (water and lipid), and thus P^s must equal 1. Hence the state defined by the equilibrium spreading pressure consists of a monolayer phase in equilibrium with excess lipid phase. The dashed line drawn as a continuation of the curve in region III is the Π–A relation for the supercompressed monolayer.

2.3.5. Binary Mixtures

For binary lipid surface mixtures, the treatment becomes more complex. The simplest case is the one treated in Section 2.2, where both lipid components are present in excess. In this case, $P^b = 3$ (two lipid phases and water), $C = 3$, and for constant temperature and pressure the phase rule may be written as $F = 1 - P^s$. Since F cannot be negative and P^s cannot be zero, $F = 0$ and $P^s = 1$. The equilibrium film will have a single surface phase characterized by a single value of surface pressure and surface composition. It was on this basis that the immiscibility of cholesterol and DPL was determined (Section 2.2).

If the restriction that both lipid components are at saturation is relaxed, the phase rule indicates that the possible number of surface phases is not unequivocally determined under all conditions. For example, consider the typical film penetration experiment where one of the lipid components is spread as a monolayer on the surface of a solution which contains a second lipid component. The conditions are set so that the surface pressure is measured at a fixed surface area, or the film area is altered to maintain the surface pressure constant. If it is assumed that both lipids are below the saturation concentrations at the experimental temperature and pressure, for $P^b = 1$ (water), $F = 3 - P^s$, and the phase rule indicates three possibilities arising from this set of conditions. P^s can be either 1 or 2, and under both conditions Π will be an independent variable and a function of the composition of the system. The third possibility is that $F = 0$ and $P^s = 3$. This implies the formation of a complex which is insoluble in the other two phases. I am not aware that this situation has as yet been observed.

These considerations apply equally well to spread mixed films, where the difficulties are further exacerbated by the problems associated with supercompression. Thus the majority of mixed film studies have been made under conditions which may have included phase separation and supercompression. The interpretations placed on many of these studies have generally been based on the assumptions that the monolayer was homogeneous and at equilibrium. In view of the theoretical uncertainty in treating these systems, another approach is required which will allow for a better definition of the system. The thermodynamics of these systems and deductions are presented in the following section.

2.4. The Film Balance Experiment: A Thermodynamic Analysis

In the preceding sections, the emphasis of the discussion has been on establishing criteria for equilibrium and the difficulties to be expected with condensed lipid monolayers. Considering all the possible difficulties entailed, it is understandable why most monolayer studies have yielded only qualitative information. It is therefore appropriate to ask what monolayer experiments can reasonably be expected to give insights into understanding the properties of membrane lipids.

The most general properties of lipid films are the intermolecular energies, since they are characteristic of each type of film structure (Gershfeld, 1970). In bulk systems, the internal energies may be evaluated from the heats of vaporization (for liquids) or sublimation (for solids). The comparable process for the lipid film—evaporation from the surface—is obviously more complex, since it involves an extra energy term which is equivalent to disrupting the interactions between lipid and water in addition to the forces which are normally present between the lipid molecules themselves. In principle, the internal energy of lipid monolayers may be evaluated by the sequence of processes diagrammed in Fig. 3.

Surface evaporation, CA in Fig. 3, the transfer of lipid molecules from a condensed surface state, involves changes in lipid–lipid interactions as well as lipid–water interactions which are characteristic of that state. The energetics of this process are extremely difficult to measure, though evidence

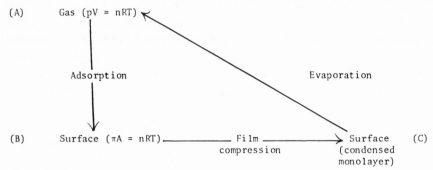

Fig. 3. **Cyclic process for evaluation of intermolecular energies in monolayers.** (A) is the (bulk) ideal gas state of lipid above the water surface; (B) and (C) are the ideal gas and condensed film states, respectively.

for evaporation from some condensed monolayers has been reported (see Section 2.1). However, with equilibrium processes, alternative pathways may be chosen which begin and terminate in the same states as surface evaporation. Hence, surface evaporation may be replaced by the equivalent processes of adsorption (AB) plus film compression (BC) shown in Fig. 3. Adsorption is assumed to involve only changes in hydration, i.e., that the film molecules behave as an ideal gas on the surface. Film compression, on the other hand, involves changes in both lipid–lipid interaction and possible changes in lipid–water interactions. It is the film compression process which is measured in the film balance experiment.

The cyclic processes outlined in Fig. 3 illustrate an important aspect of the film balance experiment, namely, that it is impossible to measure the contribution of the aqueous subphase to the properties of the monolayer without considering also the adsorption or evaporation processes. The formalism for evaluation of the energies outlined in Fig. 3 will be presented in some detail for single- and binary-component systems.

2.4.1. Single-Component Systems: Energetics

The energetics of lipid films are evaluated by measuring the temperature dependence of the free energy of film compression (ΔF_c) from the Π–A isotherms; all other thermodynamic functions may be calculated from these data.

The surface pressure Π is defined by the relation

$$\Pi = -\left(\frac{\partial F}{\partial A}\right)_{T,p,v,n} \tag{4}$$

where F is the Helmholtz free energy of the system, A is the area of the surface, with the temperature (T), pressure (p), volume (v), and number of moles of lipid (n) all held constant. In the film balance experiment, the area of the surface is usually expressed in terms of the number of moles of lipid present in the area A; thus $A' = A/n$, where A' is the area per mole of lipid in the surface. If $n = 1$, then $dA' = dA$; equation (4) may then be rewritten with $A = A'$. In all the equations which follow, the prime notation will be dropped and A will be taken to mean the area per mole or area per molecule of lipid in the surface; which definition will be apparent from the usage.

Integration of equation (4) between the limits of A and A_i yields

$$F - F_i = - \int_{A_i}^{A} \Pi \, dA \qquad (5)$$

If A_i is chosen such that $\Pi A = kT$ (where A refers to cm^2/molecule) in the region of the isotherm where the film behaves formally as an ideal gas in two dimensions, then by definition the work involved in forming the monolayer state which exists at area A will be the free energy of film compression

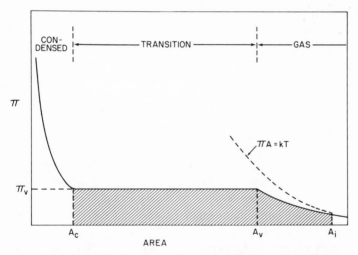

Fig. 4. Schematic Π–A isotherm of a single-component lipid monolayer. A_i is the area where the film behaves as an ideal gas ($\Pi A = kT$); A_v is the average area/molecule of gaseous film in equilibrium with condensed film of area/molecule A_c. The magnitude of the shaded area is ΔF_c. From Gershfeld (1970), reproduced by permission.

$\Delta F_c = F - F_i$. The integration of equation (5) may be performed graphically from the Π–A data shown in Fig. 4. The corresponding heats (ΔH_c) and entropies (ΔS_c) of film compression are evaluated from the temperature dependence of ΔF_c (Gershfeld and Pagano, 1972a).

2.4.2. Binary Mixtures

When two lipids are present in the surface, the system may be treated formally as in the case of bulk mixtures (Goodrich, 1957; Gaines, 1966; Pagano and Gershfeld, 1972a). However, given the uncertainties introduced by the possibility of supercompression and immiscibility (see Section 2.2), it is best to set the conditions of the experiment so as to restrict the number of possible phases in the system and simplify the analysis.

If two condensed lipid films (e.g., liquid–condensed or liquid–expanded) are used to form a mixture at a surface concentration where each component separately is in a single homogeneous state, the phase rule is written $F = 3 - P^s$ for $C = 3$ (water and components A and B), $P^b = 1$ (water), at constant T and p. The same indefiniteness with respect to identifying the number of surface phases exists as for the film penetration system described earlier (Section 2.3).

However, if the lipids are mixed in surface concentrations where separately each component is in equilibrium with its own surface vapor, i.e., the transition region (see Fig. 4), the phase relations are greatly simplified. Thus at constant T and p, $F = 3 - P^s$. If the components are miscible in the condensed state, $P^s = 2$ (condensed mixed film AB, and surface vapor AB), and $F = 1$; hence Π will vary with the composition of the surface. If A and B are not miscible, $P^s = 3$ (the two immiscible condensed phases, plus the surface vapor) and $F = 0$, and Π will be constant and independent of the surface composition.

The data may be presented in a way analogous to that for bulk systems, as in the mixing of two liquids in equilibrium with vapor. The surface vapor pressure is plotted as a function of the composition of the surface film where the composition may be either for the condensed phase or for the vapor phase.* The types of curves to be expected are given in Fig. 5, where X refers

---

* The composition is obtained by measuring Π_v where there is very little surface vapor, i.e., where A is low, and where most of the film is in the condensed state. If the composition of the surface vapor is desired, Π_v is measured at A very large, where most of the film molecules are in the vapor state, yet where some of the film is still in the condensed state. The Π_v/X diagrams will closely resemble those obtained by measurements of the mixing of two liquids (e.g., Glasstone, 1946).

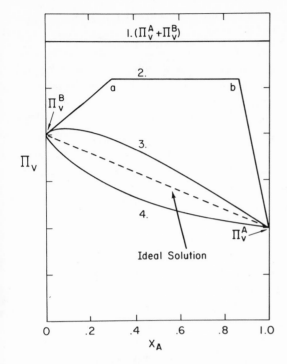

Fig. 5. Surface vapor pressure Π_v vs. mole fraction X_A for surface solution containing components A and B. The mole fraction refers to the composition of the condensed phase: Curve 1, complete immiscibility; curve 2, partial miscibility; curves 3 and 4, complete miscibility with positive (3) and negative (4) deviations from ideal solution behavior; i.e., the surface solution obeys Raoult's law. From Gershfeld (1972); reproduced by permission.

to the composition of the condensed phase(s). Curve 1 is for complete immiscibility, where the measured surface pressure will be the sum of the surface vapor pressures of the components $\Pi_v^A + \Pi_v^B$. Complete miscibility will be indicated by curves 3 and 4, for which the surface pressure varies continuously as a function of composition between the two surface vapor pressures. The dashed line indicates the behavior expected if the two components form an ideal mixture, i.e., where the surface analogue of Raoult's law is obeyed. Curve 2 indicates partial miscibility, where points a and b represent compositions of two phases which coexist between points a and b. If the overall mole fraction is varied between a and b, the relative amounts of the two phases will vary, but the mole fraction of each phase will remain constant. These results are in accordance with the predictions of the phase rule.

It is of interest to note that the curves presented in Fig. 5 have direct analogies with the bulk properties of liquid mixtures and that the interactions between the components are reflected by the type of deviations of the Π–X curves from Raoult's law, i.e., positive or negative deviations. To evaluate the excess energies of mixing requires the following relations:

The free energy of mixing of two lipid components 1 and 2 is

$$\Delta F^m = \Delta F_c^{(12)} - x_1 \Delta F_c^{(1)} - x_2 \Delta F_c^{(2)} + RT(x_1 \ln x_1 + x_2 \ln x_2) \quad (6)$$

where $\Delta F_c^{(1)}$, $\Delta F_c^{(2)}$, and $\Delta F_c^{(12)}$ are the work of film compression for the pure components and mixed films, and x_1 and x_2 are the mole fractions of the components. The last term of equation (6) represents the ideal free energy of mixing, and the excess energies may be obtained by subtracting the ideal mixing term. From the temperature dependence of ΔF^m, all the thermodynamic parameters for mixing energies may be obtained. The same precautions discussed earlier for evaluation of film compression, energies still apply; hence it is important to obtain the isotherms for all three systems (1, 2, and 12) into the gaseous region. In earlier studies (Goodrich, 1957), excess thermodynamic function measurements from lipid surface film mixtures were obtained which ignored this region of the isotherm. As an example of the magnitude of the errors involved, the following free energies were obtained for a mixture of stearic and oleic acids calculated to test the importance of using the extended isotherm. These data are taken from a recent report (Pagano and Gershfeld, 1972b) and are presented in Table II. It is clear from these data that the errors can be very large indeed.

Table II. Dependence on Film Area of Thermodynamic Parameters for Mixed Films of Stearic and Oleic Acids ($T = 298°K$)

Parameter[a]	A: 20–45 Å²/molecule (cal/mole)	A: 50–10⁵ Å²/molecule ($\pm$ 25 cal/mole)
$\Delta F_c^{(1)}$	+550	+2135
$\Delta F_c^{(2)}$	+ 80	+1130
$\Delta F_c^{(12)}$	+390	+1770
ΔF_{xs}	+120	+ 240
ΔF^m	−275	− 160

From Pagano and Gershfeld (1972b).
[a] Component 1, oleic acid; component 2, stearic acid; mixture 12, stearic acid plus oleic acid (3:2). $\Delta F_{xs} = \Delta F^m - \Delta F^{ideal}$.

2.4.3. Film Penetration

Phase homogeneity is difficult to establish in film penetration studies, where one lipid component is a spread film and a second component is added to the subphase. This can be demonstrated by analysis of this experiment in terms of the phase rule. The phase rule for this system becomes, for constant temperature and pressure, $F = 4 - P^s - P^b$. Thus if $F = 1$ (variable Π), $P^s + P^b = 3$. Since one phase is the subsolution and there is at least one surface phase, the third phase may be either bulk lipid or another surface phase. However, the conditions of the penetration experiment can be fixed to give useful information. The Gibbs equation for this system is written for components 1 and 2

$$d\Pi = \Gamma_1 \, d\mu_1 + \Gamma_2 \, d\mu_2 \qquad (7)$$

If the conditions are set so that the spread film is at saturation with excess crystals of the lipid present, the surface pressure will be Π_e, the equilibrium spreading pressure; with variable composition of the second component in the subsolution, the Gibbs equation becomes

$$d\Pi_e = \Gamma_2 \, d\mu_2 \qquad (8)$$

since $d\mu_1 = 0$ (the chemical potential of the spread film remains constant if it is in equilibrium with excess phase of lipid, component 1). By plotting Π_e as a function of the chemical activity of component 2 in the subsolution, the slope of the curve is Γ_2. Alternatively, if Γ_2 is measured directly by using radiotracers (see Section 3.3), the chemical potential of component 2 is obtained.

This section has analyzed monolayer systems which are amenable to theoretical treatment and which provide a rigorous experimental framework for examining questions of structure and interactions between different components in surfaces. Classical thermodynamics considers general relations between macroscopic properties and is not directly concerned with the molecular properties of the system. However, it is the predictive aspects of the molecular properties of model systems which are of most interest. This aspect is best treated by statistical mechanics, which deals with the problem of relating macroscopic properties with intermolecular forces. The next section will summarize some results of statistical mechanics pertinent for treating thermodynamic data from lipid films obtained with the film balance.

2.5. Molecular Thermodynamics of Lipid Monolayers

Evaluation of the intermolecular energies in lipid monolayers will involve lipid–lipid and lipid–water interactions. As indicated earlier (Section 2.4), the contribution of the underlying water to the structure of the film cannot be measured rigorously from the film balance experiment. In principle, these energies can be obtained from the energetics of surface evaporation (Fig. 3), but there are obvious experimental difficulties associated with the measurements—the vapor pressure of spread films is extremely low (see Section 2.1). From the standpoint of what can be measured in the film balance experiment, it is clear that the film compression process represents relative changes in the water contribution from the initial to the final compressed state of the film.

Despite this shortcoming of the film balance experiment, useful information about the structure of the films may be obtained provided a suitable method can be devised which will permit the separation of the lipid–water interactions from the lipid–lipid interactions. In the following sections, the method of extracting the hydrocarbon interactions is described. It is based on a comparison of the properties of the lipid region of the monolayer with the properties of bulk hydrocarbons. The assumption is made that the phase transitions in the lipid region of the monolayer will follow the same dependence on the length of the hydrocarbon chain as do analogous properties in bulk. The first section summarizes the bulk properties of hydrocarbons and their dependence on n, the number of -CH_2 groups in the aliphatic chain, and the second section extends the analysis to properties of lipid films.

2.5.1. Bulk Properties of Hydrocarbons: Dependence on n

For hydrocarbon liquids and solids, the processes of interest are those for evaporation and sublimation. The analysis which follows is taken from Huggins (1939), who developed relations for the dependence of the energetics of various processes as a function of n, the number of -CH_2 groups in the molecule; for illustrative purposes, only those functions for the evaporation of hydrocarbons will be presented, and the reader may refer to the original sources for the analysis of the other systems.

The heat of evaporation of a hydrocarbon liquid is the difference between the heat contents of the gas and liquid states. For the gas, the heat content has contributions from the various modes of translational, vibrational, and rotational motions, the covalent bond energies, and the van der Waals cohesion energy. For the liquid, the sum of the translational, rotational, and vibrational energies plus the covalent bond energies will be the same as for the gas. Accordingly, the major difference between the liquid and gas is the van der Waals force contribution. Langmuir (1925, 1926) first calculated the evaporation energy for an aliphatic compound by assuming that the configuration of the gas molecule was that of a sphere with the interior CH_2 groups in the same environment as in the bulk liquid but with the exterior groups exposed to air and having the same potential energy as at the oil–air interface. According to this model, the difference in energy between the gaseous and liquid states is the contribution from the surface energy of the hydrocarbon sphere in the gas.

For the dependence of the energy of evaporation on n, Huggins utilized Langmuir's model and deduced an expression for the heat content of the gas which depends on an $n^{2/3}$ relation. As the hydrocarbon chain length increases, the surface area of the gas particle and hence the potential energy increase as $n^{2/3}$. Using experimental data, at 298 °K a plot of ΔH_v, the heat of evaporation, against $n^{2/3}$ was linear and passed through the origin. Thus at 298 °K, $\Delta H_v = 2.56 \times 10^3 n^{2/3}$. Huggins attributes the deviations from the theoretical slope at $n < 10$ as due to the nonspherical shape of the lower molecular weight compounds.

There has recently been some criticism of the use of the $n^{2/3}$ dependence on theoretical grounds (Meyer and Stec, 1971), but the alternative is an expression linear in n but including an additional empirical constant. This constant is impossible to evaluate in spread monolayers, and hence this linear dependence of ΔH_v on n is not useful for the monolayers.

The dependence of the entropy of vaporization on n follows similar arguments. The entropy of a long-chain compound will be the sum of four terms: (1) the external vibrations of the molecules as a whole, (2) the position and orientation of each molecule as a whole, (3) orientations of each bond relative to each other and from rotation about the bonds, and (4) vibrations of the atoms in each molecule relative to each other. The same general considerations apply to both the liquid and gaseous states, and by summing all contributions Huggins finds at 298 °K

$$\Delta S_v = 6 + 0.3n + 8 \ln n \qquad (9)$$

where the constants are derived empirically from experimental data. The Gibbs free energy of vaporization, $\Delta G_v = \Delta H_v - T\,\Delta S_v$, with the terms for the entropy and enthalpy, becomes

$$\Delta G_v = 2.56 \times 10^3 n^{2/3} - T(6 + 0.3n + 8 \ln n) \qquad (10)$$

This equation is satisfied by the experimental vapor pressure data for $n \geqslant 10$, where most of the deviations from this relation at small n are due to the enthalpy term (Huggins, 1939). An equivalent expression for the free energy may be obtained by plotting the values for this function against n; the result is a linear expression:

$$\Delta G_v{}^{(298)} = -830 + 454n \qquad (11)$$

This equation holds when n equals 10–20.

The application of these empirical expressions for free energies, heats, and entropies of vaporization of hydrocarbons to the lipid region of condensed lipid films will be analyzed in the section which follows.

2.5.2. Lipid Film Properties

For the film properties, we use the Helmholtz free energy F, and assume the $\Delta F \approx \Delta G$. To extract the contributions of the interactions in the hydrocarbon region of the lipid film, it is necessary to assume that the polar group contribution is additive and that a physically meaningful relation exists between the properties of the film and the number of -CH$_2$ groups in the chain. In support of the first assumption, it has been demonstrated (Gershfeld, 1970) that the hydrocarbon and polar group contributions to the free energy of film compression are separable and additive. When ΔF_c was plotted as a function of n for various homologous lipids, the data were grouped onto two series of lines with constant slopes. Each slope was a function of the physical state of the lipid in the film and not of the particular polar group. Thus the hydrocarbon contributions to liquid–condensed and liquid–expanded films may be represented by a term nf, where f is the contribution of each -CH$_2$ group to the total free energy of compression.

The second assumption may be tested by considering, with Langmuir (1925, 1926), that the hydrocarbon region of the condensed lipid films may be treated as a liquid hydrocarbon with an internal energy that may be measured from the surface evaporation process (Fig. 3). The free energy of film compression for this hypothetical hydrocarbon liquid may be obtained from the equivalent processes $-\Delta F_c^h = \Delta F_{ads}^h + \Delta F_v^h$, where superscript h indicates that the particular process is for the hydrocarbon contribution. As an approximation of the adsorption process, we choose those data for the adsorption of hydrocarbon gases on water (Jones and Ottewill, 1955) as shown in Fig. 6; the data indicate that the adsorption process for the hypothetical lipid moiety may be represented by a function nf_{ads}^h. For the vaporization process, the discussion in section 2.5.1 (equation 11) suggests that ΔF_v^h may be represented by a similar function nf_v^h. Thus one can expect $nf_c^h = -n(f_v^h + f_{ads}^h)$, which verifies the conclusion reached previously (Gershfeld, 1970) that a linear relation between ΔF_c and n is physically meaningful.

A similar argument applies for the hydrocarbon contribution to the heat of film compression (ΔH_c^h). The dependence on n for the heats of ad-

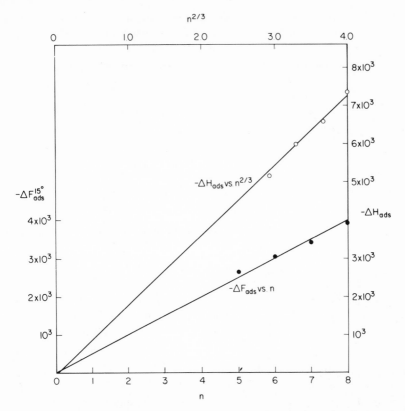

Fig. 6. Dependence on hydrocarbon chain length (n) of thermodynamics of adsorption of hydrocarbon gases on water. Data from Jones and Ottewill (1955). ($\bigcirc$, ΔH_{ads} vs $n^{2/3}$ $\bullet$, ΔF_{ads} vs n.

sorption of hydrocarbon gases on water (Jones and Ottewill, 1955) obeys the relation $\Delta H_{ads}^{h} = kn^{2/3}$ as seen in Fig. 6, and for the evaporation of hydrocarbon liquids $\Delta H_{v}^{h} = k'n^{2/3}$ (Section 2.5.1). Thus one can expect that for the film compression process $\Delta H_{c}^{h} = k_c n^{2/3}$.

Actual calculations for hydrocarbon energies of lipid films are presented in Table III, taken from a recent publication (Gershfeld and Pagano, 1972a). The values listed are for the sum of film compression and adsorption, which is equivalent to the process of evaporation; for comparison, the values for the comparable processes with bulk hydrocarbons (solid and liquid) are also presented. The energies for the hydrocarbon region of the condensed monolayers are about the same as for a hydrocarbon liquid.

It is appropriate to summarize the limitations of the approach which

Table III. Comparison of the Thermodynamics of Transition for Bulk and Monolayer Hydrocarbons at 15°C for 1 Mole of a C_{15}-Hydrocarbon ($n = 15$)

Transition process from bulk gas to:	ΔH_{trans} (kcal/mole)	ΔS_{trans} (e.u.)
Bulk[a]		
a. Solid	−34.1	−58.6
b. Liquid	−21.6	−32.6
Monolayer		
a. Solid	−24.9	
b. Liquid–condensed	−24.7	−28 ± 3
c. Liquid–expanded	−20.1	−30 ± 3

[a] Calculated from the relations given by Huggins (1939).

has been outlined. One of the major difficulties is verification of the dependence of the thermodynamic parameters of film compression on the hydrocarbon chain length (n). Unfortunately, it is not presently practicable to verify the dependence independently with the film balance. The major restriction is that for the dependence to be tested rigorously all of the members of the homologous series must be in the same physical state so that a proper comparison may be made. For many of the lipid systems available, the range of carbon chain lengths where the physical state remains the same is usually restricted to only a few carbons. For example, with $n = 13–15$ the films are liquid–expanded, and for larger n the films are liquid–condensed. However, with the longer chain length compounds the surface vapor pressures become increasingly smaller to the point where they cannot be measured with any certainty. This sets the lower limit to about 18–20 carbons, and a range of about 3–5 carbon atoms, which is too restrictive to make any test of the dependence on n significant.

3. EXPERIMENTAL METHODS

The principal experimental parameters for measuring the thermodynamic properties of lipid films are the surface pressure Π and the surface concentration Γ.* For the former, a film balance has been developed with a

---

* "Surface concentration" is used synonymously with the correct expression "surface excess concentration."

sensitivity of 0.001 dyn/cm (Gershfeld *et al.*, 1970; Pagano and Gershfeld, 1972*b*), while for the latter various techniques are available depending on the nature of the system. The sections which follow will describe the essential experimental details which are necessary for the measurement of these parameters. Each of the techniques discussed has been directly assessed in our laboratory.

3.1. The Film Balance

The film balance consists of a sensing element for measuring the surface pressure, a trough for containing the subsolution, and barriers for controlling the total surface area in which the spread film is contained. The general techniques for measuring surface tensions are not considered here, since it is the "insoluble" films which are of primary interest. However, if absolute values of surface tension are required, these techniques are well described elsewhere (Harkins, 1945; Gaines, 1966). Each of the elements of the film balance will be considered in turn, since there are special problems associated with each.

The two principal methods for measuring surface pressures are the horizontal float system, developed by Langmuir (1917), and the plate

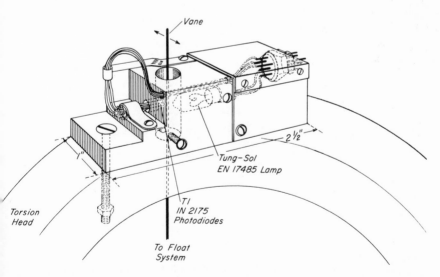

Fig. 7. Scale drawing of sensor fixed in position to torsion head of commercial film balance. From Gershfeld *et al.* (1970); reproduced by permission.

method (Wilhelmy, 1863), which was extensively studied by Harkins and Anderson (1937). The major requirement for the sensing element is that surface pressures of the order of 0.001 dyn/cm be routinely measured. While the plate method can provide that sensitivity, the method suffers from possible artifacts arising from the irreproducibility of the wetting of the plate (Gaines, 1966). The error in Π may be an order of magnitude greater than the required sensitivity, which eliminates the plate technique as a method for the measurement of intermolecular energies. However, the plate method is adequate for measuring surface pressures which exceed 1 dyn/cm and is particularly useful for measuring interfacial pressures at liquid–liquid interfaces.

The technique of choice for routine measurement of surface pressures in the low-pressure range of the isotherm is the horizontal float system,

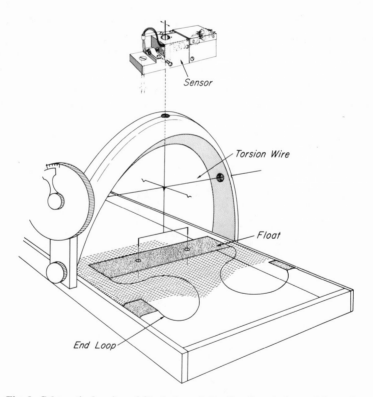

Fig. 8. Schematic drawing of film balance indicating the relative positions of the sensing element and the float system. From Pagano and Gershfeld (1972b); reproduced by permission.

which does not suffer from the wetting artifact of the plate method. A system recently developed for this purpose (Pagano and Gershfeld, 1972*b*) consists of a sensor which monitors the displacement of the float. The sensor consists of an incandescent source of illumination and two semiconductor photodiodes (Fig. 8). A vane, attached to the indicator pointer of the horizontal float assembly, is positioned between the photodiodes and the incandescent bulbs. The photodiodes are sensitive only to light which is nearly parallel to their axes and within a cylinder of radius which is considerably smaller than the radius of the diodes. The width of the vane is slightly larger than the center-to-center diode distance. The vane initially is positioned so that it blocks most of the light which would ordinarily reach the diodes, but when there is a slight movement of the vane due to the displacement of the float a relatively large decrease in the light which reaches one diode is matched by an equivalent increase in the amount of light which reaches the other diode. The current in each diode is proportional to the light which reaches the diode so that the difference in current at the juncture of the diodes may be amplified and calibrated in terms of the displacement of the float; this displacement is directly proportional to the surface pressure. A more elaborate description of the sensor is presented elsewhere (Gershfeld *et al.*, 1970).

The float system is comprised of a strip of Teflon, 0.0025 cm thick and 0.5 cm wide; the length varies depending on the width of the trough. The attachment of the float to the sensor is as indicated in Fig. 8. Most of the difficulty inherent in making the system essentially frictionless is the attachment of the strip to the sides of the trough via the end loops, which are generally made of flexible material. One major source of difficulty results from the fact that this end loop must traverse the region of the meniscus and in so doing is distorted to the point where film leakage is likely. To obviate this problem, thin Teflon strips are attached to the trough walls where the length of the strip is sufficient to extend beyond the curvature of meniscus with its extremity lying just up to the edge of the float (see diagram, Fig. 9). Attached to the fixed strip and the float is the end loop, made of surgical silk which has been lightly paraffined. The end loop coils naturally in the plane of the water surface, and, with small surface pressures, assumes a semicircular shape.

The trough and barriers have been fabricated from a variety of materials: brass, steel, aluminum, polymethyl methacrylate, Teflon, and glass. Teflon has the virtue of being readily cleansed and at the same time chemically inert. The major difficulty with its use as a barrier is its tendency

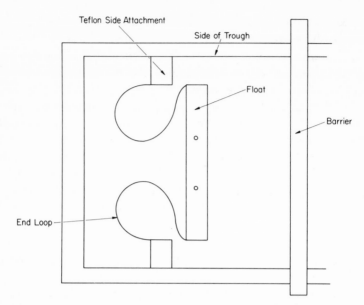

Fig. 9. Diagram of the float system indicating the positions of the float, side attachments, and end loops.

to undergo cold flow, and therefore it will tend to bow after prolonged use in one position. This difficulty is avoided by bracing the barrier with a rigid strip of metal. Convenient dimensions for barriers are 15 by 1.0 by 1.5 cm. The barrier should extend about 2 cm beyond the edges of the trough.

At least five barriers should be available, two on each side of the float for sweeping surface contaminants from the surface, and one to vary the surface area in which the spread film is confined.

Control of temperature and methods for depositing monolayers have been extensively discussed (Harkins, 1945; Gaines, 1966) and need not be repeated here.

3.2. Evaluation of Γ Using Radiotracers

The importance of directly obtaining values for the surface excess concentration of adsorbed lipid components has been stressed earlier in the section on film penetration. In this section, the experimental methods for evaluating Γ and their limitations will be discussed.

The total radioactivity (R_x) emanating from the surface of a solution which contains a surface-active labeled compound can be written

$$R_x = R_{s,x} + R_{b,x} \tag{12}$$

where the subscripts s and b refer to the surface and bulk contributions of the labeled compound. Three separate measurements are required to evaluate Γ_x, as described in the following sequence:

Let subscripts x, 1, and 2 represent different compounds labeled with the same isotope (e.g., ^{14}C) with (1) a water-insoluble compound which forms a spread film, (2) a water-soluble compound which is not surface active, and (x) the unknown compound which is both water soluble and surface active.

The first measurement is the determination of the relative specific activities (σ_x/σ_1), and (σ_x/σ_2). To obtain these values, the general relation

$$R = wK'\sigma \tag{13}$$

is used, where w is the amount of the compound, K' is the detection co-efficient (which includes the geometry of the detecting system, counting efficiency, back- and self-scattering factors, etc.), and σ is the specific activity. Thus, from a measurement of the radioactivities R of the three compounds under identical physical conditions so that K' is identical for each compound, and with known amounts of each compound w, relative specific activities may be calculated.

The second measurement is $R_{b,2}$, for the non-surface active (2) compound. $R_{b,x}$ follows from the relation

$$R_{b,x} = (C_x/C_2)(\sigma_x/\sigma_2)R_{b,2} \tag{14}$$

where C is the concentration in solution of the indicated compound. Determination of $R_{b,2}$ gives $R_{b,x}$ since all other factors are known. By subtracting $R_{b,x}$ from R_x, $R_{s,x}$ is determined [equation (12)].

To determine Γ_x, a third measurement must be made which is the radioactivity of the spread film (1) as a function of its surface concentration. The relation for evaluation of Γ_x is

$$\Gamma_x = (\Gamma_1/R_{s,1})(\sigma_1/\sigma_x)R_{s,x} \tag{15}$$

Evaluation of $\Gamma_1/R_{s,1}$ is obtained by spreading a given amount of the compound 1 on the water surface and measuring the radioactivity of the surface film. Since all other factors are known, Γ_x is then determined.

The sensitivity of the method depends on several factors, as seen from the relation

$$R_{s,x} = \frac{1}{1 + (1/m)(C/\Gamma)} \tag{16}$$

where m is the mass absorption coefficient (cm^2/g) for β-rays in solution. The highest accuracies can be expected for small values of C/Γ and for large values of m. Values of m for a group of useful isotopes are given in Table IV. The most useful isotope is ^{3}H, for which m is about 50 times larger than for ^{14}C or ^{35}S.

Table IV. Radioisotopes for Use in Monolayers

Isotope	Halflife	Radiation	Radiation energy (MeV)	Approximate maximum range in water	m^a (cm^2/g)
^{3}H	12.3 years	β^-	0.018	6 μ	1.66×10^3
^{14}C	5760 years	β^-	0.155	300 μ	33
^{35}S	87.2 days	β^-	0.167	340 μ	30
^{36}Cl	3.03×10^5 years	β^-	0.714	$\sim$2 mm	5
^{32}P	14.2 days	β^-	1.71	$\sim$1 cm	1
^{45}Ca	165 days	β^-	0.25	650 μ	15
^{131}I	8.04 days	β^-	0.61	$\sim$2 mm	5
		γ	0.36		

From Gershfeld (1972).
^{a} Mass absorption coefficient.

For most studies with isotopes, excluding tritium, ordinary end-window Geiger–Mueller tubes of gas-flow detectors may be used. For the special problem of counting tritium, the range in air and the thickness of the window do not permit the counting of this isotope by a Geiger–Mueller tube. However, recent efforts have shown that extremely thin windows may be made of plastic materials, which then allows for the detection of tritium at the air–water surface (Frommer and Miller, 1965) by gas-flow counters.

A more sophisticated device has been developed by Muramatsu et al. (1967, 1968); this technique utilizes essentially a gas-flow windowless counter in which the aqueous surface of the trough is entirely enclosed, with the detector, in an environment which is filled with a humidified hydrocarbon gas. The detector is a plastic sheet which has incorporated a phosphor that scintillates upon being bombarded by radiation. The scintillations are transmitted to photomultiplier tubes by a prismatic light-guide. The major advantage of this system is that the detector can be incorporated into commercially available scintillation counters which permit the simultaneous recording of radiation from different isotopes; thus double-labeling experiments can be performed on the same film.

Finally, the radioactive purity of the compounds used must be verified by thin-layer chromatography or an equivalent technique. Clearly, if the labeled material is an impurity in the compound, the results are for the impurity.

3.3. Preparation and Use of Materials for the Film Balance

Lipids can be obtained with varying degrees of purity from various commercial sources. With the advent of gas thin-layer chromatography, characterization of the lipids with respect to purity is now routine. In addition to methods of characterization, there are the surface properties of Π_e and Π_v, which can be routinely measured.

It is virtually impossible to prepare "pure" water for use with the film balance. The very nature of the use of the trough, with its large exposed water surface to air and the trough, precludes the elimination of all contaminants. The best situation to aim for is the preparation of routinely reproducible water for which the experimental variables are not sensitive to the inevitable contaminants which will appear. Two of the most serious contaminants are surface-active organic material and heavy metal ions.

Water which has been distilled from quartz and stored in quartz flasks is reasonably certain to be free of heavy metal contaminants. However, when electrolytes are added to the water, heavy metal ionic contamination is surely present, and for ionic films this can be a serious problem. To eliminate this source of contamination, "sweeping" the surface with an ionized insoluble film will remove most of the ionic heavy metal contamination. At pH 5–8, alternate spreading and sweeping of four films of dioctadecylphosphate films are sufficient to remove Ca^{2+} from a half liter of 10^{-7} M solution (Pak and Gershfeld, 1964). At pHs greater than 8, behenic acid films are equally efficient (Spink, 1963).

For the elimination of surface-active contaminants, continual sweeping of the surface of the solution in the trough must be continued until no change in film pressure can be detected when the surface area is reduced to 1 % of the full area of the trough. This process may take as long as several hours.

One should avoid the use of charcoal, EDTA, and ion-exchange resins in the treatment of the water, since their use has been shown to introduce contaminants into the water (Pak and Gershfeld, 1964).

The most convenient method for introducing a given amount of lipid to the surface is by depositing the lipid from a volatile solvent. The controversy over the effects of the residual solvent on the monolayer properties has been associated with the solubility of the solvent in water. However, for film studies at the very low surface pressure region, recent studies indicate that very little residue remains if sufficient time is allowed for evaporation (Table V). Thus any of the solvents listed in Table V will be useful and lead to reproducible results.

Table V. Effect of Spreading Solvents on Surface Pressures

Solvent	Boiling point ($^\circ$C)	Solubility in H_2O (g/1000 g H_2O, 25°C)	Time for Π to reach 0.001 dyn/cm (min)
n-Hexane	69	0.01	1
Benzene	80	1.8	5
Methanol	65	∞	30

Another potential source of artifact arises from the possibility that the lipid will form stable aggregates in the spreading solvent which may persist even after spreading on the water surface (Pagano and Gershfeld, 1972b). Where this problem is suspected, use of several different types of solvent systems is recommended.

Pipettes suitable for dispensing these solutions have been adequately described (Harkins, 1945). Obviously, the limitations will be on the precision of the delivered volumes, and careful regard must be paid to the calibration of the pipette from this standpoint. A standard isotherm for a stable compound such as octadecanol may be used to verify the volumetric calibration.

3.4. Other Techniques

Other techniques which are not the source of primary information (one in which the theoretical basis is not established) are surface potentials and surface viscosity. The nature of these techniques has been extensively discussed (Gaines, 1966) and shall not be considered in detail here. However, it is useful to note that these techniques, while only providing indirect information about the structural properties of lipid films, are capable of providing a measure of the change in state of a film. For example, the surface potential of a homologous series will be the same, provided all the members of the series are in the same physical state (Adam and Harding, 1933). Moreover, the surface viscosity changes when a film passes through a phase change (Harkins, 1945, 1952).

Another worthwhile approach is use of spectrometers which utilize multiple reflections through the spread film and an integrator for the exit beam (Tweet, 1963; Gaines *et al.*, 1964). Attenuated total reflectance utilizing built-up films on crystal surfaces has also given useful results, but the structures of the built-up films are more like crystal structures than the spread-film structure. Thus the relevance to monolayer studies is somewhat limited.

More recently, self-exchange diffusion coefficients of lipids in surface films have been measured. The initial results suggest that diffusion coefficients in films may be of the same order of magnitude as in free solution (Good and Schechter, 1972). However, precautions must be taken to assure that there is only one phase present in the film. For the systems which have been examined, the possibility of multiphase behavior is indicated, which limits the significance of the results. This is an important new technique for the study of monolayer properties, and in principle should lead to significant information about the structure of lipid films.

Many nonequilibrium properties of monolayers have also been examined, and were not discussed in this review. Their omission here was not intended to diminish the importance of many of the nonequilibrium properties but rather to limit the subject matter to a reasonable size. However, a brief mention of some of these properties may be of interest. In particular, the rates of film spreading (Davies and Rideal, 1961), transport of gases through films (Blank, 1972), and the water drag phenomenon (Pak and Gershfeld, 1967) provide another view of the monolayer and involve hydrodynamic properties of the monolayer system. Additional references to some of these topics may be found elsewhere (Davies and Rideal, 1961).

ACKNOWLEDGMENTS

The author wishes to express his appreciation to the following for granting permission to reproduce material from their copyrighted publications:

Academic Press, Inc., for Fig. 4 and 8; The American Institute of Physics, for Fig. 7; Marcel Dekker, Inc., for Fig. 5.

4. REFERENCES

Adam, N. K., 1941, "The Physics and Chemistry of Surfaces," 3rd ed., Oxford University Press, London.

Adam, N. K., and Harding, J. D., 1933, The structure of surface films. Part XX. Surface potential measurements on nitrites, *Proc. Roy. Soc. (Lond.)* **A143**:104.

Adam, N. K., and Jessop, G., 1926, The structure of thin films. Part VII. Critical evaporation phenomena at low compressions, *Proc. Roy. Soc. (Lond.)* **A110**:423.

Blank, M., 1972, Measurement of monolayer permeability, *in* "Techniques of Surface and Colloid Chemistry and Physics," Vol. 1 (R. J. Good, R. R. Stromberg, and R. L. Patrick, eds.), pp. 42–84, M. Dekker, New York.

Brooks, J. H., and Alexander, A. E., 1960, Losses by evaporation and solution from monolayers of long chain aliphatic alcohols, *3rd Internat. Congr. Surface Activity (Cologne)* **2**:196.

Brooks, J. H., and Alexander, A. E., 1962a, Spreading and collapse phenomena in the fatty alcohol series, *J. Phys. Chem.* **66**:1851.

Brooks, J. H., and Alexander, A. E., 1962b, The spreading behavior and crystalline phases of fatty alcohols, *in* "Retardation of Evaporation by Monolayers" (V. K. LaMer, ed.), p. 245, Academic Press, New York.

Crisp, D. J., 1949, A two dimensional phase rule. II. Some applications of a two dimensional phase rule for a single interface, *in* "Surface Chemistry," p. 23, Supplemental Research, London.

Davies, J. T., and Rideal, E. K., 1961, "Interfacial Phenomena," Academic Press, New York.

Defay, R., 1932, Thesis, University of Brussels.

Defay, R., Prigogine, I., Bellemans, A., and Everett, D. H., 1966, "Surface Tension and Adsorption," pp. 74–78, Wiley, New York.

Devaux, H., 1913, Oil films on water and on mercury, *Smithsonian Inst. Ann. Rep.* **1913**:261.

Findlay, A., 1951, "The Phase Rule," 9th ed., Dover, New York.

Frommer, M. A., and Miller, I. R., 1965, A method for measuring the adsorption of tritium labeled compounds and its application in the investigation of surface activity of DNA, *J. Colloid Interf. Sci.* **21**:245.

Gaines, G. L., Jr., 1966, "Insoluble Monolayers at Liquid–Gas Interfaces," Interscience, New York.

Gaines, G. L., Jr., Tweet, A. G., and Bellamy, W. D., 1964, Interaction between chlorophyll a and vitamin K_1 in monomolecular films, *J. Chem. Phys.* **42**:2193.

Gershfeld, N. L., 1968, Cohesive forces in monomolecular films at an air–water interface, *Advan. Chem. Ser.* **84**:115.

Gershfeld, N. L., 1970, Intermolecular energies in condensed lipid monolayers on water, *J. Colloid Interf. Sci.* **32**:167.

Gershfeld, N. L., 1972, Film balance and the evaluation of intermolecular energies in monolayers, *in* "Techniques of Surface and Colloid Chemistry and Physics," Vol. 1 (R. J. Good, R. R. Stromberg, and R. L. Patrick, eds.), pp. 1–39, M. Dekker, New York.

Gershfeld, N. L., and Pagano, R. E., 1972*a*, Physical chemistry of lipid films at the air–water interface. I. Intermolecular energies in single-component lipid films, *J. Phys. Chem.* **76**:1231.

Gershfeld, N. L., and Pagano, R. E., 1972*b*, Physical chemistry of lipid films at the air–water interface. III. The condensing effect of cholesterol. A critical examination of mixed-film studies, *J. Phys. Chem.* **76**:1244.

Gershfeld, N. L., and Patlak, C. S., 1966, Activity coefficients of monomolecular films from desorption studies, *J. Phys. Chem.* **70**:286.

Gershfeld, N. L., Pagano, R. E., Friauf, W. S., and Fuhrer, J., 1970, Millidyne sensor for the Langmuir film balance, *Rev. Sci. Inst.* **41**:1356.

Gibbs, J. W., 1948, "Collected Works," Vol. 1, p. 219ff, Yale University Press, New Haven.

Glasstone, S., 1946, "Textbook of Physical Chemistry," 2nd ed., Chapter 10, Van Nostrand, Princeton, N.J.

Good, P. A., and Schechter, R. S., 1972, Surface diffusion in monolayers, *J. Colloid Interf. Sci.* **40**:90.

Goodrich, F. C., 1957, Molecular interaction in mixed monolayers, *in* "Proceedings of the Second International Congress on Surface Activity," Vol. 1, p. 85, Butterworths, London.

Hardy, W. B., 1912, The tension of composite fluid surfaces and the mechanical stability of films of fluid, *Proc. Roy. Soc. (Lond.)* **A86**:610.

Harkins, W. D., 1945, Determination of properties of monolayers and duplex films, *in* "Physical Methods of Organic Chemistry," Vol. 1 (A. Weissberger, ed.), p. 211, Interscience, New York.

Harkins, W. D., 1952, "The Physical Chemistry of Surface Films," Rheinhold, New York.

Harkins, W. D., and Anderson, T. F., 1937, A simple, accurate film balance of the vertical type for biological and chemical work, and a theoretical and experimental comparison with the horizontal type, *J. Am. Chem. Soc.* **59**:2189.

Heikkila, R. E., Kwong, C. N., and Cornwell, D. G., 1970, Stability of fatty acid monolayers and the relationship between equilibrium spreading pressure, phase transformations, and polymorphic crystal forms, *J. Lipid Res.* **11**:190.

Huggins, M. L., 1939, Certain properties of long-chain compounds as functions of chain length, *J. Phys. Chem.* **43**:1083.

Jones, D. C., and Ottewill, R. H., 1955, Adsorption of insoluble vapors on water surfaces. Part II, *J. Chem. Soc.* **1955**:4076.

Langmuir, I., 1917, The constitution and fundamental properties of solids and liquids. II. Liquids, *J. Am. Chem. Soc.* **39**:1848.

Langmuir, I., 1925, The distribution and orientation of molecules, *in* "Third Colloid Symposium Monograph," p. 48, Chemical Catalogue Co., New York.

Langmuir, I., 1926, The effects of molecular dissymmetry on properties of matter, *in* "Colloid Chemistry," Vol. 1 (J. Alexander, ed.), p. 525, Chemical Catalogue Co., New York.

Langmuir, I., 1939, Molecular layers, *Proc. Roy. Soc. (Lond.)* **A170**:1.

Lyons, C. G., and Rideal, E. K., 1929, On the stability of unimolecular films. Part I. The conditions of equilibrium, *Proc. Roy. Soc. (Lond.)* **A194:**322.

Mansfield, W. W., 1959, The influence of monolayers on evaporation from water storages. III. The action of wind, wave, and dust upon monolayers, *Austral. J. Appl. Sci.* **10:**73.

Meyer, E. F., and Stec, K. S., 1971, Evidence against energetically favored coiling of vapor-phase paraffins up to *n*-tetracasane, *J. Am. Chem. Soc.* **93:**5451.

Muramatsu, M., and Ohno, T., 1971, A radiotracer study on hydrolysis of methyl-C^{14} palmutate in insoluble monolayers, *J. Colloid Interf. Sci.* **34:**469.

Muramatsu, M., Tokunaga, N., and Koyano, A., 1967, End-window counting of tritium radioactivity by scintillation phosphor combined with prismatic light guide, *Nucl. Inst. Meth.* **52:**148.

Muramatsu, M., Shingematsu, A., and Tokunaga, N., 1968, An improved scintillation probe for efficient counting of low energy beta rays, *Nucl. Inst. Meth.* **55:**249.

Pagano, R. E., and Gershfeld, N. L., 1972*a*, Physical chemistry of lipid films at the air–water interface. II. Binary lipid mixtures. The principles governing miscibility of lipids in surfaces, *J. Phys. Chem.* **76:**1238.

Pagano, R. E., and Gershfeld, N. L., 1972*b*, A millidyne balance for measuring inter-molecular energies in lipid films, *J. Colloid Interf. Sci.* **41:**311.

Pak, C. Y. C., and Gershfeld, N. L., 1964, The detection and removal of trace calcium from water for charged monolayer studies, *J. Colloid Interf. Sci.* **19:**831.

Pak, C. Y. C., and Gershfeld, N. L., 1967, Steroid hormones and monolayers, *Nature* **214:**888.

Patlak, C. S., and Gershfeld, N. L., 1967, A theoretical treatment for the kinetics of mono-layer desorption from interfaces, *J. Colloid Interf. Sci.* **25:**503.

Rabinovitch, W., Robertson, R. F., and Mason, S. G., 1960, Relaxation of surface pressure and collapse of unimolecular films of stearic acid, *Canad. J. Chem.* **38:**1881.

Rideal, E. K., 1930, "An Introduction to Surface Chemistry," 2nd ed. Cambridge University Press, London.

Roylance, A., and Jones, T. G., 1960, Monolayer formation from hexadecanol crystals, *3rd Internat. Congr. Surface Activity (Cologne)* **2:**123.

Schulman, J. H., and Hughes, A. H., 1935, Mixed unimolecular films, *Biochem. J.* **29:**1243.

Spink, J. A., 1963, Ionization of monolayers of fatty acids from C-14 to C-18, *J. Colloid Sci.* **18:**512.

TerMinassian-Saraga, L., 1955, Study of adsorption and desorption at liquid surfaces. III. Desorption of monomolecular layers, *J. Chem. Phys.* **52:**181.

Tweet, A. G., 1963, Spectrophotometer for optical studies of ultra-thin films, *Rev. Sci. Inst.* **34:**1413.

van Deenen, L. L. M., Hontsmuller, U. M. T., de Haas, G. H., and Mulder, E., 1962, Monomolecular layers of synthetic phosphatides, *J. Pharm. Pharmacol.* **14:**429.

Wilhelmy, L., 1863, Über die Abhängigkeit dur Capillaritäts-Constanten des Alkohols von Substanz und Gestalt des denetzten festen Körpers, *Ann. Phys.* **119:**177.

Chapter 3

Circular Dichroism and Absorption Studies on Biomembranes

D. W. URRY AND M. M. LONG

Laboratory of Molecular Biophysics
University of Alabama Medical Center
Birmingham, Alabama

1. INTRODUCTION

Application of circular dichroism and absorption to the study of the structure and function of biomembranes is a complex effort because of the multitude of interactions that require simultaneous consideration and the distorted spectra which result from the particulate nature of membranes. Justification of the effort derives from recognition that many processes fundamental to biological systems, for example, neuronal activity, mitochondrial energy transformation, active and passive transport, hormonal control of cellular activity, and muscle contraction, are membrane-associated phenomena. In contrast to the difficulties, however, a substantial advantage of these optical methods is that membranes in active, functional states may be observed.

1.1. On the Complexity of Biomembrane Spectra

Perhaps an optical study of biomembranes can be placed in perspective by progressing stepwise from simpler solution situations. Spectroscopic studies of molecules in solution characteristically utilize, as a reference state, concentrations where interactions between molecules are negligible. As a relatively simple situation, consider *p*-cresol, the chromophoric side-chain of tyrosine. This is a single chromophoric moiety. In dilute solutions, the absorption spectrum is dependent on interactions with the solvent. The absorption spectrum dependence on solvent or environment in itself is a complex problem concerned with the refractive index and dipole moment of the solvent and with the dipole moments of the ground and excited states of the chromophoric moiety (Bayliss and McRac, 1954; Yanari and Bovey, 1960). Aggregation of identical chromophores such as p-cresol results in splitting and shifting of the absorption bands due to excitation resonance interactions (exciton interactions) between identical transitions (Kasha, 1963) and also results in hyper- or hypochromism of individual absorption bands due to dispersion force interactions, i.e., a coupling of an electronic transition in one chromophoric moiety with nonidentical transitions in an adjacent moiety (Rhodes, 1961).

1.1.1. A Molecule with Several Chromophoric Moieties

An increasingly complex situation would be a molecule containing two or more different chromophoric moieties. An example which has relevance to membranes would be a heme peptide containing the heme and peptide chromophores. Absorbances of the heme are perturbed by dispersion force interactions with transitions in the peptide chromophores, and another additional factor is the coordinations of the fifth and sixth positions of the heme iron. Also, the heme transitions become optically active, largely because of dispersion force interactions with peptide transitions, and the peptide transitions in a reciprocal manner derive part of their optical activity from coupling with heme transitions (Tinoco, 1962; Urry, 1965, 1967a). Aggregation of heme peptides becomes more complex because of the number of dispersion force and exciton-type interactions that require consideration. The situation in the aggregate can be somewhat simplified by noting the overriding intensity of the heme absorbances and neglecting coupling with peptides. This and other approximations lead to information on the relative orientation and distance between heme moieties in the aggregates (Urry, 1967b; Urry and Pettegrew, 1967). By treating this as a model system, it provides a basis for considering the proximity of multiple hemes in membranous preparations of cytochrome oxidase (Urry et al., 1967; Urry and van Gelder, 1968) and the problem of heme–heme interaction in hemoglobin (Urry and Pettegrew, 1967).

1.1.2. High Polymers

The large polymers of biological systems, e.g., proteins and polypeptides, derive distinctive spectral characteristics from intramolecular interactions. These polymers often assume specific conformations, each with characteristic optical rotation and absorption properties. Thus we have come to recognize α-helical patterns (Holzwarth and Doty, 1965; Grosjean and Tari, 1964) and β-pleated-sheet patterns (Sarkar and Doty, 1966; Iizuka and Yang, 1966). Here questions arise concerning the interactions which are responsible for the patterns, the contribution of the invariable backbone, and the effect of the variable side-chains. This problem is best resolved by obtaining polypeptides in the characteristic conformations but without side-chain chromophores. The polypeptides poly-L-alanine and poly-L-serine best satisfy this requirement and provide spectra characteristic of the α-helix and the antiparallel β-pleated sheet without the problem of side-chain chromophores (Quadrifoglio and Urry, 1968a,b). Accordingly, the circular dichroism and absorption spectra of these polymers unambiguously show the contribution of the backbone conformation and as such become fundamental reference states for the α-helical and β-pleated-sheet patterns (Quadrifoglio and Urry, 1968a,b). Addition of side-chain chromophores, caused by side-chain chromophore–peptide chromophore and side-chain chromophore–side-chain chromophore interactions, can modify the fundamental spectra obtained on poly-L-alanine and poly-L-serine. It has become common practice to resolve an optical rotatory dispersion or a circular dichroism curve of a protein into contributions due to the characteristic patterns. In this approach, one must decide on a set of reference states; one then assumes no other conformations exist, neglects contributions due to variable side-chain chromophores and chromophores in prosthetic groups, and neglects interactions between segments or domains identifiable with specific conformations. It is apparent that the assumptions and difficulties increase with the size of the molecular system under consideration.

1.1.3. Aggregation of High Polymers and Biomembranes

On aggregation of polymers, the additional interactions are intermolecular. Also, aggregation can cause changes in the intramolecular interactions due to induced conformational changes. These effects are qualitatively similar to preceding considerations. When the molar particle weight is of the order of many millions, however, a new set of problems ensue which markedly distort the optical spectra. This chapter presents an analysis of these spectral distortions as they have been proposed and detailed in this laboratory (Urry, 1972a); it compares efforts to calculate a model aggregate system, poly-L-glutamic acid; it discusses the presence or absence of distortions at 222 nm in biomembranes; it outlines approximate, but practical, methods of achieving corrected spectral data; and it provides an outline of the compli-

cated analysis of suspension data once the data have been corrected for
particulate distortions.

The biomembrane—being a complex mixture of highly absorbing and
highly optically active proteins, of lipids with generally low optical activity
and absorbance, and of abundant other specialized quantities, such as
substrates, cofactors, heme moieties, channels, and carriers—requires much
caution in data interpretation. Recognizing the limitations of data interpretation in terms of detailed membrane structure, work is noted which
demonstrates the value of empirical correlations in comparing different
membrane systems as well as the same membrane system in different metabolic states and which attempts to obtain relevant solubilized systems and to
achieve meaningful reconstitution.

2. DISTORTIONS IN THE ABSORPTION AND CIRCULAR DICHROISM SPECTRA OF BIOMEMBRANES

2.1. Absorption Effects

When light passes through a suspension, photons are scattered in directions
other than toward the phototube. This decrease in number of photons
reaching the phototube is not distinguished instrumentally from true
absorbance by the sample. That component of the measured absorbance
which arises due to light scattering is designated A_S.

A second effect was described by Duysens (1956) as an absorption
flattening; it leads to decreased absorbance in regions of absorption bands
of molecules comprising the aggregate. As the absorbance of an increment
of path is proportional to the intensity of light entering the increment,
anything which causes an abrupt drop in light intensity just prior to entering
the path increment will result in a decrease in absorbed photons in the
increment. In more descriptive terms, it can be said that an absorbing
particle casts a shadow obscuring the chromophores in the beam path behind
the particle. This results in a flattening of the absorption band. Duysens
defined an absorption-flattening quotient (Q_A) as the ratio of suspension
absorbance (A_{susp}) to the correct solution absorbance (A_{soln}); i.e.,

$$Q_A = \frac{A_{susp}}{A_{soln}} \tag{1}$$

Since the suspension absorbance is more complicated, we refer to a flattened absorbance

$$A_F = Q_A A_{\text{soln}} \tag{2}$$

where A_F is not necessarily A_{susp}.

The added complication which we have proposed is referred to as a concentration obscuring due to light scattering. As our state of reference is the molecularly dispersed solution, an effort is made to correct the suspension data in a manner removing the light-scattering and absorption-flattening effects. In a solution at a specified wavelength, there is a probability that a given photon will be absorbed, or, in other words, a certain fraction of the photons are absorbed. In a suspension at a given wavelength, a fraction of the light is scattered. We take the position that in order to correct for the total scattering effect, it should be appreciated that a fraction of the scattered light would have been absorbed by a molecularly dispersed solution with the same concentration of absorbers, and we approximate that the fraction of the scattered light that would have been absorbed by the suspension is the same as the fraction of light which would have been absorbed in the solution. For a solution, one writes

$$A = - \log \frac{I}{I_0} \tag{3}$$

where I is the emerging beam intensity and I_0 is the initial beam intensity. If I_A is the absorbed intensity, that is, $I = I_0 - I_A$, then

$$A = - \log \frac{I_0 - I_A}{I_0} \tag{4}$$

and

$$A = - \log (1 - X_A) \tag{5}$$

where

$$X_A = \frac{I_A}{I_0} \tag{6}$$

Therefore, X_A is the fraction of light absorbed, or it is the probability that a photon is absorbed. An analogous expression can be written for the measured absorbance due to scatter:

$$A_S = - \log (1 - X_S) \tag{7}$$

where

$$X_S = \frac{I_S}{I_0} \tag{8}$$

with I_S being the intensity of light scattered by the suspension in directions other than toward the phototube. It is then approximated that the obscured absorbance due to light scattering (A_{OBSC}) is given by

$$A_{\text{OBSC}} = -\log(1 - X_A X_S) \tag{9}$$

$$A_{\text{OBSC}} = -\log(1 - X_{\text{OBSC}}) \tag{10}$$

or that

$$X_{\text{OBSC}} = X_A X_S \tag{11}$$

Accordingly, we write the absorbance of a suspension as

$$A_{\text{susp}} = A_F - A_{\text{OBSC}} + A_S \tag{12}$$

Obviously, for each value of X_A and X_S, a value can be obtained for A_{OBSC}. The question is whether to use A_{soln} or A_F to calculate X_A. When A_F is used, there is a prime placed on the fraction of absorbed light; i.e.,

$$A_F = -\log(1 - X'_A) \tag{13}$$

Equation (9) is rewritten

$$A_O = -\log(1 - X'_A X_S) \tag{14}$$

and equation (12) becomes

$$A_{\text{susp}} = A_F - A_O + A_S \tag{15}$$

A plot of A_O (or A_{OBSC}) vs. A_S is given in Fig. 1 for various values of A_F (or A_{soln}). An interesting feature seen in Fig. 1 is that as A_F (or A_{soln}) becomes greater than 1, the obscured absorbance A_O (or A_{OBSC}) approaches A_S in magnitude and equation (15) (or 12) reduces to

$$A_{\text{susp}} = A_F \tag{16}$$

This explains why the Duysens' treatment fits satisfactorily at an absorption maximum. Also, it provides a means of approximating Q_A at the absorption maximum. Another point to note is that A_{OBSC} is a function of both absorption and scattering. A_{OBSC} is a cross-term and corrects for the somewhat artificial separation of absorption and scattering.

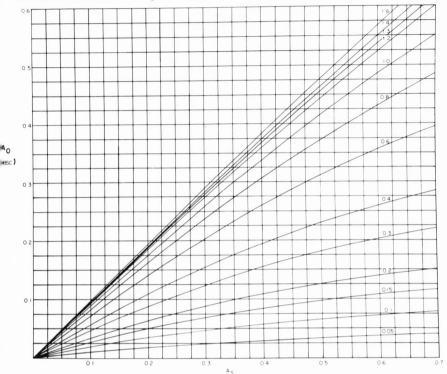

Fig. 1. Relationship between a given A_S and the corresponding A_{OBSC} for various true absorbance values. It is obtained by plotting equation (9). X_A of equation (9) and the corresponding absorbance can either be for the solution absorbance or for the flattened absorbance of the suspension. In the latter case, X_A becomes X_A' in equation (13), and A_{OBSC} becomes A_O as in equation (14). Note that when A_F (or A_{soln}) becomes greater than 1.5, A_S is approximately the same magnitude as A_O and equation (16) applies. From Urry (1972b).

2.2. Circular Dichroism: Differential Absorption Effects

Circular dichroism is a difference absorbance measurement between left and right circularly polarized light, i.e., $A_L - A_R$. This difference quantity, when appropriately converted to a molar quantity, is called "ellipticity" $[\theta]$.

$$[\theta] = \frac{3300}{C_0 l} (A_L - A_R) \tag{17}$$

where C_0 is the analytical concentration of the unit of interest, usually referring to a specific chromophore, and l is the path length in centimeters. For a suspension, we write

$$[\theta]_{\text{susp}} = \frac{3300}{C_0 l} (A_L - A_R)_{\text{susp}} \tag{18}$$

such that

$$[\theta]_{\text{susp}} = Q_E [\theta] \tag{19}$$

where Q_E, the ellipticity-distorting quotient, is given by

$$Q_E = \frac{(A_L - A_R)_{\text{susp}}}{(A_L - A_R)} \tag{20}$$

Utilizing equations (15) and (2), the numerator of equation (20) is written

$$(A_L - A_R)_{\text{susp}} = (Q_{AL}A_L - Q_{AR}A_R) - (A_{OL} - A_{OR}) + (A_{SL} - A_{SR}) \tag{21}$$

where the first term in parentheses on the right side of the equation is the differential absorption-flattening expression, the second is the absorption-obscuring term, and the last is the differential light-scattering term.

It has been shown when $X_{SL} = X_{SR}$ that to a good approximation (Urry, 1972a)*

$$\frac{Q_{AL}A_L - Q_{AR}A_R}{A_L - A_R} \simeq Q_A^2 \tag{22}$$

and

$$\frac{A_{OL} - A_{OR}}{A_L - A_R} \simeq Q_A^2 (1 - e^{-\sigma}) \tag{23}$$

where

$$\sigma = \frac{A_O}{A_{\text{soln}}} \tag{24}$$

such that the ellipticity-distorting quotient becomes

$$Q_E^{A_{SL}=A_{SR}} \simeq Q_A^2 e^{-\sigma} \tag{25}$$

When neglecting the differential scatter effect, the corrected suspension value becomes:

---

* The differential absorption-flattening effect, $(Q_{AL}A_L - Q_{AR}A_R)/A_L - A_R)$, early formulated by this laboratory (Urry, 1970) is equivalent (Urry, 1972a) to the averaged flattening quotient for circular dichroism designated as Q_B by Gordon and Holzwarth (1971). The first literature reference utilizing Q_{AL} and Q_{AR} was Urry et al. (1968).

$$[\theta]^{A_{SL}=A_{SR}} \simeq \frac{[\theta]_{\text{susp}}}{Q_A^2\, e^{-\sigma}} \qquad (26)$$

On adding the contribution due to differential scatter (equations 18 and 22), one obtains

$$[\theta]_{\text{corr}} \simeq \frac{[\theta]_{\text{susp}} - (3300/C_0 l)\, (A_{SL} - A_{SR})\, 10^{-A_F}}{Q_A^2\, e^{-\sigma}} \qquad (27)$$

2.3. Instrumental Artifacts

Instrumental artifacts can compound the distortions introduced in the CD spectra of particulate systems. These artifacts should be recognized and eliminated. Any deviation from the free-beam baseline which is induced by an optically inactive, absorbing solution is artifactual. If the instrument is challenged with such a solution and the absolute value of the molar ellipticity is greater than zero, it can be safely assumed that the instrument has introduced the error. Frequently, a similar deviation from baseline occurs with optically inactive suspensions which are light scattering. This artifact can be recognized by placing a suspension of colloidal silica or an opaque window, such as a ground-quartz window, in the beam and checking the response relative to the free-beam baseline. If the two do not coincide, an instrumental artifact is present which can be corrected by adequate alignment of the pockel cell, lens, and phototube. Both instrumental artifact tests are required to substantiate the accuracy of the circular dichroism instrument used in membrane studies.

3. ON CALCULATIONS OF THE PARTICULATE POLY-L-GLUTAMIC ACID MODEL SYSTEM

Poly-L-glutamic acid (PGA) provides a satisfactory model for considering the artifacts or distortions produced in absorption and circular dichroism spectra of suspensions. This is because a molecularly dispersed reference state can be obtained which already contains the intermolecular interactions introduced on aggregation and because in the aggregated state the polymer can be shown to retain its solution conformation. More specifically, near pH 3.9 less than 5% of the carboxylic acid side-chains are ionized (Urry,

1968). With this degree of ionization, the helical rods of PGA can aggregate to a limited extent. The number of molecules in the aggregate is about ten (Tomimatsu *et al.*, 1966), giving a particle molar weight of the order of 1×10^6. Accordingly, the particle size is too small to give rise to the particulate distortions, but it is sufficiently large to contain most of the effects of intermolecular coupling of transitions. Furthermore, the ionization effect is essentially complete, and, significantly, infrared studies on D_2O solutions and suspensions indicate no change of backbone conformation on aggregation (Urry *et al.* 1970*a*).

Two different laboratories have attempted calculations of the distortions expected in suspensions of PGA. This laboratory has used the *ad hoc* phenomenological analysis outlined in Section 2 of this chapter. It is an analysis which treats the beam of light more as a stream of discrete photons with probabilities of absorption and scatter than as electromagnetic waves. Gordon (1972) has employed the Mie scattering theory, which utilizes the classical electrodynamic treatment of light and which enjoys the respect due a theory that has proven useful for more than 60 years (Mie, 1908).

The experimental work on PGA aggregates and the two efforts at calculation are contained in Fig. 2. In Fig. 2A, the experimental particulate curves *b* and *c* are seen progressively to dampen and red shift as the extent of aggregation increases. In Fig. 2B, the calculated curves of Gordon also exhibit the tendency to dampen and red shift as particle size increases from $R \sim 0$ to $R = 0.03\,\mu$ to $R = 0.1\,\mu$. The red shifting, however, is too great and the dampening too little. This is particularly evident by comparing the magnitude of the shift in the crossover with the extent of dampening or decreased magnitude of the initial negative band. The ratio $\Delta\lambda$ (nm) over $\Delta(A_L - A_R)$ is about 3 for curve *c* in Fig. 2A, whereas it is 24 for curve *c* of Fig. 2B. What this indicates is an excessive calculated contribution of differential light scattering. It is also apparent by looking at the positive bands that this extremum is red shifting far faster than it is dampening. Similarly, comparison of the experimental absorption curves (Urry *et al.*, 1970*a*) with those calculated by the Mie scattering theory (Gordon, 1972) shows the calculated spectra to contain excessive light scattering. While these calculations are of limited value in a quantitative way, they do lend the weight of Mie scattering theory to the presence of distortions and specifically to the concept of differential light scattering of the left and right circularly polarized beams. Support of the presence of distorted spectra for membranes and the significance of differential light scattering is important, as these proposed distortions (Urry and Ji, 1968; Urry and Krivacic, 1970) had

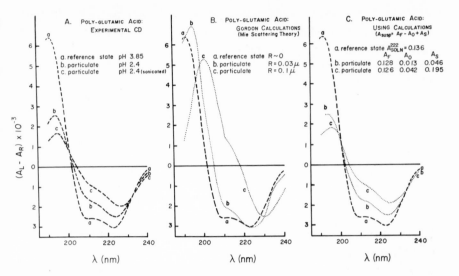

Fig. 2. Poly-L-glutamic acid (PGA) suspensions. The dashed curves are experimental data (from Urry *et al.*, 1970*a*), and the dotted curves are calculations of spectra for particulate systems. A: The reference state is taken at *p*H 3.85, where the protonation is essentially complete (5 % ionized) and where aggregation is to the extent of about ten helical rods. The aggregate molar weight is about 1×10^6. (This degree of aggregation is sufficient to introduce the effects of interactions between helices but insufficient to result in particulate distortions.) On lowering the *p*H to 2.4, the particulate distortions are substantial (curve *b*). Sonication at *p*H 2.4 leads to larger aggregates and greater distortion (curve *c*). From Urry (1972*a*) and (1973*a*). B: Using the data of Urry *et al.* (1970*a*), Gordon (1972) has calculated the form of the distortions to be expected on the basis of Mie scattering theory. Curve *b* is calculated for PGA particles with 0.03 μ radii, and curve *c* is for PGA particles with 0.1 μ radii. Comparison of curves *b* and *c* with experimental curves *b* and *c* in A shows a gross similarity. C: Curves *b* and *c* were calculated (Urry, 1972*b*) using the *ad hoc* phenomenological analysis outlined in Section 2 of the text. At 222 nm, one can neglect differential light scattering for predominantly α-helical systems; i.e., $[\theta]_{susp} = [\theta]_{soln} Q_A^2 \cdot Q_\sigma$. With the values for curve *b*, of $Q_A = 0.136$ and $Q_\sigma = e^{-\sigma}$ with $\sigma = 0.013/0.136$, giving $Q_A^2 = 0.88$ and $Q_\sigma = 0.91$, we have a correction factor of 0.80. Comparison of the calculated curves *b* and *c* with the experimental curves *b* and *c* in A shows how satisfactorily the distorted spectra can be calculated. This provides the basis for the approach to biomembranes.

been vigorously contested (Ottaway and Wetlaufer, 1970; Gordon *et al.*, 1969).

In Fig. 2C are given the calculated curves using the approach outlined in Section 2 herein and treated in greater detail previously (Urry, 1972*a*). Curves *b* and *c* of Fig. 2C virtually superimpose on curves *b* and *c* of Fig. 2A.

The capacity of this approach to calculate distorted circular dichroism spectra of the particulate PGA model system demonstrates that we have a relatively simple and practical means of correcting distorted spectra. When applied to biomembranes, this approach utilizes simultaneous determination of absorption and circular dichroism spectra and a suitable molecularly dispersed state of the membrane. The criteria required to identify a suitable molecularly dispersed state, called a "pseudo reference state", are discussed below. If a satisfactory pseudo reference state cannot be found, meaningful correction can still be obtained near $[\theta]_{222}$, which is a wavelength of interest in relating the membrane CD pattern to those of characteristic polypeptide conformations.

4. ON THE PRESENCE OR ABSENCE OF ARTIFACT AT 222 nm IN THE CD OF BIOMEMBRANES

While there is now general acceptance of the presence of distortions in the circular dichroism of biomembranes and even substantial agreement on the nature or source of the distortions, there remain efforts to identify specific wavelengths which contain little or no artifact. For example, Glaser and Singer (1971) state, "In particular it has been shown that the value of $[\theta]_{222}$ is not significantly influenced by optical artifacts, and that in the intact (erythrocyte) membranes, the protein is on the average about 40% in the right handed α-helical conformations." And Litman (1972) more recently reiterates, "$[\theta]_{222}$ appears to be less sensitive to these optical artifacts than shorter wavelength ellipticities and reliable estimates of helicity in unsonicated preparations can be obtained using this value". Before treating the question of distortions at 222 nm, it should be realized that even in the best of circumstances, e.g., with a pure globular protein, calculated values of helical content require reservation. The situation is, of course, more tenuous in the study of biomembranes with their complexity and specialized functions.

Two of the optical artifacts, absorption flattening and differential light scattering, have only limited contribution at 222 nm. Since absorbance by membranes at 222 nm is less than one-fourth the absorbance at the 190 nm maximum, membranes are commonly studied under conditions where the absorption flattening at 222 nm is only about 10%. As the differential light scattering effect is similar in shape to the optical rotatory dispersion curve which for α-helix and β-conformations crosses zero near 222 nm, the contribution of differential light scattering can be negligible at this wavelength. Consideration of only these two sources of spectral distortion does not explain the decrease in magnitude of $[\theta]_{222}$ in poly-L-glutamic acid suspensions (see Fig. 2). In the calculations represented in Fig. 2C, the dampening is caused by concentration obscuring due to light scattering which is contained in the factor $e^{-\sigma}$ (see equation 27). (The quotient Q_σ is also used for $e^{-\sigma}$ or for expansions of $e^{-\sigma}$.) The experimental results on PGA dramatically show $[\theta]_{222}$ to increase in magnitude with decrease in particle size. What remains is to demonstrate similar pattern changes for biomembranes.

Sequential sonication of mitochondria in KCl results in a stepwise increase in $[\theta]_{222}$ to a point where the increase in magnitude is some 400% (see Fig. 3). Sonication of the $(Na^+ + K^+)$ ATPase from brain (Skou, 1965;

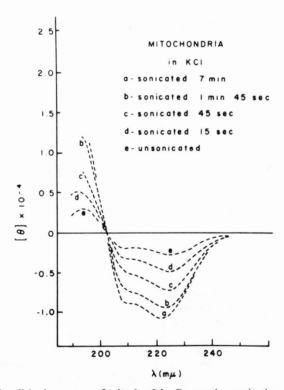

Fig. 3. Circular dichroism curves of mitochondria. Progressive sonication results in progressive increases in magnitude of ellipticity. The concentration of protein was 1.88 mg/ml and the path length 0.213 mm. Note the progressive red shift of the long wavelength extremum as one approaches the intact mitochondria, i.e., as the particle size increases. Most apparent are the dramatic increases in ellipticity at 222 nm. From Urry et al. (1971b).

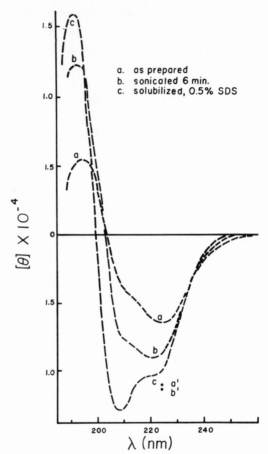

Fig. 4. **Sonication of (Na^+ + K^+) ATPase from brain.** Curve *a* is the CD spectrum of sodium iodide treated pig-brain microsomes containing the sodium- and potassium-dependent transport ATPase. Curve *b* was obtained after the membrane preparation had been sonicated on ice for 6 min. The effect of sonication was to decrease particle size with a negligible solubilization as tested by measuring the CD of the supernatant following centrifugation of the sonicated sample. Sodium dodecylsulfate at 0.5% solubilized the NaI microsomes, as seen in curve *c*. The protein concentration was 0.62 mg/ml and the path length 0.23 mm. Points *a'* and *b'* are the corrected ellipticities at 224 nm for curves *a* and *b*, respectively. That the corrected points for curves with very different extents of distortion are almost the same provides confidence in the corrected values.

Long *et al.*, 1973), as seen in Fig. 4, also results in stepwise increases in ellipticity magnitude at 222 nm as well as at shorter wavelengths. Unless sonication converts large amounts of protein into conformations with the α-helical-type CD pattern (which, though unlikely, would be very interesting), it must be concluded that the ellipticity of biomembranes at 222 nm, as is the case with PGA aggregates, is highly subject to the particulate distortions. Also, it should be appreciated that sonication can only go so far in removing the distortions. The remainder of the distortions need to be removed by some formalism which has been adequately tested on a model system.

A membrane system which shows little increase in magnitude at 222 nm on sonication is the freshly prepared red blood cell ghost. This does not mean that no dampening exists at 222 nm; rather, it means that the extent of distortion cannot be decreased by sonication. If the erythrocyte ghost preparation is stored frozen and then thawed and reexamined, large dampening can be observed which is alleviated by sonication until limiting ellipticities are obtained which are similar to those of the fresh preparation.

5. APPROXIMATE CORRECTIONS

5.1. Pseudo Reference State Approach

Before the optical rotation properties of a membrane in its different functional states can be compared with the purpose of assessing conformational effects and, indeed, before such data from a series of dissimilar membranes can be contrasted, the inherent distortions must be corrected. An empirical approach to improve distorted ellipticity data is the pseudo reference state method, in which the absorption parameters of a membrane suspension and its appropriate solution are used to minimize the dampening and red shift typically found in the CD spectra of membranous systems. The validity of this approach is based on studies utilizing synthetic polypeptides whose states of aggregation can be experimentally controlled (Urry *et al.*, 1970*a*).

When a membrane is solubilized and its molecularly dispersed state meets a stringent set of criteria, this soluble system is defined as the pseudo reference state (Urry, 1972*a*). The absorption values of this membrane solution and suspension, when collected simultaneously with the CD spectra, provide the means to calculate the correction factors in equation (26), that is, Q_A and $e^{-\sigma}$ or Q_σ. When the solution molar rotation is also obtained, all the parameters in equation (27) are calculable.

To test the appropriateness of a particular solubilized membrane preparation, an initial calculation of the corrected molar ellipticity, $[\theta]_{C1}^{224}$, is made at 224 nm by using equation (26) and the absorbance at 224 for any solubilized state of the membrane. The reason for choosing 224 nm is three-fold. At this wavelength, the optical rotatory dispersion curve is near zero (the differential scatter of the right and left circularly polarized beams reduces to zero, i.e., $A_{SL} = A_{SR}$), and equation (26) becomes a valid approximation. Also, when the membrane protein changes conformation during its transition from a matrix-bound to a freely soluble macromolecule, absorbance changes due to hyper- or hypochromism are small at 224 nm. Finally, Q_A^{224} can at times be approximated by unity.

For a membrane solution to qualify as a pseudo reference state (prs), its observed molar ellipticity at 224 nm should be the same as or preferably somewhat greater than that corrected value obtained with the following:

$$[\theta]_{C1}^{224} = \frac{[\theta]_{susp}^{224}}{Q_{A^2}^{224} \, Q_{\sigma}^{224}} \tag{28}$$

Q_A^{224} is assumed to be 1—a safe assumption if the crossover point of the suspension and solution absorbances is far from 224. If not, Q_A^{224} must be calculated from equations (29) and (30) and Table I:

$$Q_A^{192} \simeq \frac{A_{susp}^{192}}{A_{soln}^{192}} \tag{29}$$

$$A_P(\lambda_1) = \frac{A_{soln\,(\lambda_1)}}{A_{soln\,(\lambda_2)}} A_P(\lambda_2) \tag{30}$$

Table I converts the absorption through the particle $A_P(\lambda_1)$ into the corresponding $Q_A(\lambda_1)$—in this case, Q_A^{224}; Q_σ is obtained from the difference in absorbance of the suspension and membrane solution; that is,

$$A_{susp}^{224} - Q_A^{224} A_{soln}^{224} = A_S^{224} - A_{OBSC}^{224} = {}^oA_S^{224} \tag{31}$$

With the value ^{o}of A_S^{224} and the solution absorbance at 224 nm, Q_σ^{224} is read off the ordinate in Fig. 5. This value of Q_σ provides a minimal correction. If equation (32) is obeyed, then the membrane solution is an appropriate pseudo reference state.

$$[\theta]_{soln,\ observed}^{224} > [\theta]_{C1}^{192} = \frac{[\theta]_{susp}^{224}}{Q_{A^2}^{224} \, Q_{\sigma}^{224}} \tag{32}$$

$[\theta]_{C1}^{224}$ is a minimal value, commonly low by more than 10%.

To check the soluble preparation even further, $[\theta]_{corr}^{192}$ is calculated and tested to see if it agrees with equation (33).

$$[\theta]^{192}_{\text{soln, observed}} > [\theta]^{192}_{C1} = \frac{[\theta]^{192}_{\text{susp}}}{Q^{192}_{A^2}} \tag{33}$$

At 192 nm Q_σ is taken as 1 and Q_A is obtained with equation (29) (Urry, 1972a). If $[\theta]^{192}_{C1}$ is less than that observed with the dissolved system, this preparation again meets the pseudo reference state criteria. Approximately correct values at 224 nm and 192 nm are taken to imply an adequate correspondence of absorbances at all other wavelengths, allowing the entire corrected spectra to be calculated.

Calculation of the complete spectrum is straightforward, once the pseudo reference state has been found (Urry, 1972a, 1973a). First, Q^{192}_A and

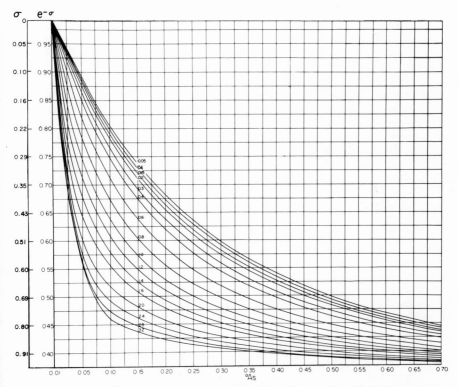

Fig. 5. A plot of $e^{-\sigma}$ vs. $^\circ As$. The latter quantity is defined by equation (34). With a given $^\circ As$, the appropriate curve (A_F) is chosen, and $e^{-\sigma}$ is directly obtained from the $e^{-\sigma}$ ordinate. The corresponding values of σ are given in a second ordinate in order that the correction for heterogeneity, schematically indicated in Fig. 6, may be included in the corrections.

Table I. Flattening Quotients and Associated Particle Absorbances

A_P	Q_A (spheres)	Q_A (vesicles)[a]	A_P	Q_A (spheres)	Q_A (vesicles)[a]
0.02	0.99	0.97	0.86	0.73	0.67
0.04	0.98	0.95	0.88	0.73	0.66
0.06	0.97	0.94	0.90	0.73	0.66
0.08	0.97	0.92	0.92	0.72	0.65
0.10	0.96	0.91	0.94	0.72	0.65
0.12	0.95	0.90	0.96	0.71	0.65
0.14	0.94	0.89	0.98	0.71	0.64
0.16·	0.94	0.88	1.00	0.70	0.64
0.18	0.93	0.87	1.02	0.70	0.63
0.20	0.92	0.86	1.04	0.69	0.63
0.22	0.92	0.85	1.06	0.69	0.62
0.24	0.91	0.85	1.08	0.68	0.62
0.26	0.90	0.84	1.10	0.68	0.62
0.28	0.90	0.83	1.12	0.68	0.61
0.30	0.89	0.82	1.14	0.67	0.61
0.32	0.88	0.81	1.16	0.67	0.61
0.34	0.88	0.81	1.18	0.66	0.60
0.36	0.87	0.80	1.20	0.66	0.60
0.38	0.87	0.79	1.22	0.66	0.59
0.40	0.86	0.79	1.24	0.65	0.59
0.42	0.85	0.78	1.26	0.65	0.59
0.44	0.85	0.77	1.28	0.64	0.58
0.46	0.84	0.77	1.30	0.64	0.58
0.48	0.84	0.76	1.32	0.64	0.58
0.50	0.83	0.76	1.34	0.63	0.57
0.52	0.82	0.75	1.36	0.63	0.57
0.54	0.82	0.75	1.38	0.62	0.57
0.56	0.81	0.74	1.40	0.62	0.56
0.58	0.81	0.73	1.42	0.62	0.56
0.60	0.80	0.73	1.44	0.61	0.56
0.62	0.80	0.72	1.46	0.61	0.55
0.64	0.79	0.72	1.48	0.61	0.55
0.66	0.79	0.71	1.50	0.60	0.55
0.68	0.78	0.71	1.52	0.60	0.54
0.70	0.78	0.70	1.54	0.59	0.54
0.72	0.77	0.70	1.56	0.59	0.54
0.74	0.76	0.69	1.58	0.59	0.53
0.76	0.76	0.69	1.60	0.58	0.53
0.78	0.75	0.68	1.62	0.58	0.53
0.80	0.75	0.68	1.64	0.58	0.52
0.82	0.74	0.68	1.66	0.57	0.52
0.84	0.74	0.67	1.68	0.57	0.52

Table I (continued)

A_P	Q_A (spheres)	Q_A (vesicles)[a]	A_P	Q_A (spheres)	Q_A (vesicles)[a]
1.70	0.57	0.51	1.86	0.54	0.49
1.72	0.56	0.51	1.88	0.54	0.49
1.74	0.56	0.51	1.90	0.54	0.48
1.76	0.56	0.50	1.92	0.53	0.48
1.78	0.56	0.50	1.94	0.53	0.48
1.80	0.55	0.50	1.96	0.53	0.48
1.82	0.55	0.50	1.98	0.53	0.47
1.84	0.55	0.49	2.00	0.52	0.47

[a] Using the formalism of Gordon and Holzwarth (1971) for vesicles to relate Q_A (vesicles) with an associated particle absorbance.

A_P^{192} are obtained from equation (29) and Table I; then equation (30) and Table I allow calculation of Q_A at every wavelength; Q_σ^λ is found with equation (34):

$$A_{\text{susp}}^\lambda - Q_A^\lambda A_{\text{prs}} = {}^\circ A_S^\lambda \tag{34}$$

and Fig. 1. Finally, the corrected ellipticity is approximated by

$$[\theta]_{\text{corr}}^\lambda = \frac{[\theta]_{\text{susp}}^\lambda}{(Q_A^\lambda)^2 \, Q_\sigma^\lambda} \tag{35}$$

It is important to note that equation (35) neglects differential scatter.

The differential-scattering artifact is eliminated when $[\theta]_{\text{corr}}^\lambda$ is obtained with equation (27), which in turn is equivalent to the following (Urry, 1973a):

$$[\theta]_{\text{corr}}^\lambda = \frac{[\theta]_{\text{susp}}^\lambda - k\,[m]_{\text{prs}} \, 10^{-A_F}}{Q_A^2 \, Q_\sigma} \tag{36}$$

where $[\theta]_{\text{soln}}^\lambda$ is the molar rotation of the membrane pseudo reference state, A_F is $Q_A \, A_{\text{soln}}^A$, and k is a constant, evaluated at one wavelength. When the molar ellipticity of the membrane solution is zero, usually around 201 nm, the difference between the suspension and solution ellipticities is largely due to differential scatter (Urry, 1973a). Therefore, equation (36) becomes

$$[\theta]_{\text{Susp}}^{201} \simeq k\,[m]_{\text{prs}}^{201} \, 10^{-Q_A \, A_{\text{soln}}} \tag{37}$$

or

$$k = \frac{[\theta]^{201}_{susp}}{[m]^{201}_{prs} \, 10^{-Q_A \, A_{soln}}} \tag{38}$$

An alternate, and perhaps more satisfactory, way to evaluate k, which does not depend on the CD of the pseudo reference state, is to utilize the change achieved by sonication of the membrane preparation. In the preceding discussions, it was noted that in the above formulation of the distortions the change in the ellipticity at $[\theta]_{222}$ on changing particle size is due to concentration obscuring arising from light scattering and that the change in crossover is due largely to the differential light scattering artifact. This means that the change of crossover and the change in magnitude at 222 nm are both a function of the extent of light scattering. When sonication can be used to decrease the degree of distortion, as in Fig. 3 and 4, a ratio $\Delta^{\lambda}_{CD}(nm)/\Delta [\theta]_{222}$ can be found which with the corrected value at 222 nm can be used to determine the expected $\Delta\lambda_{CO}$, and the required value of k is that which results in the determined $\Delta\lambda_{CO}$.

To use the above means of evaluating k, it should be verified that sonication results in a negligible amount of solubilization. This is demonstrated by centrifuging the sonication suspension and determining the CD of the supernatant. With the absorption of the suspension and the pseudo reference state, with the suspension CD data, and with the optical rotatory dispersion data on the pseudo reference state, it is possible to calculate the correction factors in equation (36) and thereby obtain a corrected spectrum.

5.2. Consideration of Membrane Heterogeneity

Our approach to correcting membrane spectra is based on the satisfactory calculations of the PGA data (see Fig. 2). Particles of PGA, however, are homogeneous polypeptide, whereas membranes are heterogeneous, being mixtures primarily of protein (or polypeptide) and lipid. If the membrane were a sea of lipid containing occasional islands of protein, then most of the scattered light would not be affected by the protein. Since the CD spectrum is in major respects due to protein, the light-scattering distortion observed in the CD would be reduced in comparison to the scattered light distortions in absorption spectrum. As the number of islands of protein become greater, i.e., as the ratio of protein to lipid increases, the light-scattering distortion in the CD becomes more relevant to that observed in absorbance. And when one is concerned with pure protein or polypeptide, then the light scattering in the absorption spectra and that in the CD spectra are integrally related. Accordingly, it would be useful when studying membranes to have a means of correcting the values of A_S obtained from the absorption spectra to quantities more directly relevant to the light-scattering effects on the CD spectra.

A convenient way to resolve the problem of membrane heterogeneity with variable protein-to-lipid ratios would be to separate the fraction of scattered light into two fractions—light scattered from protein (X_{Sp}) and light scattered from lipid (X_{Sl})—i.e.,

$$X_S = X_{Sp} + X_{Sl} \tag{39}$$

Such an approach blissfully sets aside the wave description of the light-scattering process and centers on the aspect that when light interacts with matter it does so in a manner which concentrates its interaction at a localized site, as, for example, when an electron is knocked out of a surface in the photoelectric effect.

In further efforts to obtain more meaningful membrane spectra, we are proceeding with the approach of equation (39) and introducing a correction for the volume fraction of protein (V_{fp}), where

$$X_{Sp} = (V_{fp}) X_S \tag{40}$$

V_{fp} is calculable with the weight percent of protein (W_p), the weight percent of lipid (W_l), and the densities of protein (d_p) and of lipid (d_l); i.e.,

$$V_{fp} = \frac{W_p/d_p}{W_p/d_p + W_l/d_l} \tag{41}$$

With an appropriate expression for $f(V_{fp})$, the correction for variable lipid-to-protein ratios can be introduced, as indicated in Fig. 6. Work is in progress

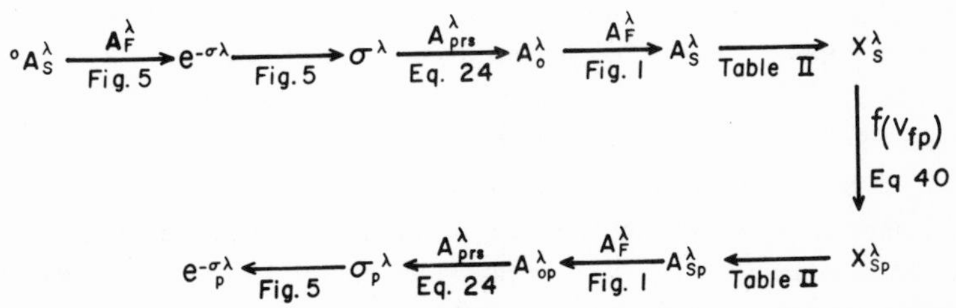

Fig. 6. Scheme to correct $e^{-\sigma}$ for the differential contribution to scattering made by the protein and the lipid of the membrane mosaic. The correct form of the function for the volume fraction of protein $f(V_{fp})$ is being studied. At present, it would likely improve the spectra to simply use V_{fp}.

to determine the proper function of V_{fp}. At the present, it is reasonable to expect that use of the factor V_{fp} alone will constitute an improvement in the corrected data.

5.3. On Alternate Sources of Absorption Data

In the preceding approach for correcting CD data on biomembranes, the absorption data are determined simultaneously on the same phototube as the CD data. It has previously been outlined (Urry, 1973a) how one may obtain simultaneous circular dichroism and absorption data by utilizing the CD photomultiplier dynode voltage. Starting at a low setting, the dynode voltage can be calibrated by plotting dynode voltage against absorbance values. With this plot, dynode voltage values can be converted to absorbance values. Absorbance curves are obtained for the reference cell and for the sample cell, and the net absorbance of the sample is obtained by the difference. This is a most useful approach, particularly when automated and when the lamp is stable between reference and sample runs. When carried out by hand, it can be tedious and the values obtained at low absorbances, e.g., near 224 nm, may be of limited accuracy. Any means which reduces the light-scattering distortions such as an end-on phototube would improve the correction by decreasing the dependence on low absorbance values.

Simultaneous absorption and CD measurement is an exact means of being assured that the same scattered light reaches the phototube in the two measurements, i.e., that the factors obtained from the absorption data are relevant to the CD data. With the accessory available, it is possible to compare absorption spectra obtained in this way with spectra obtained on standard spectrophotometers, specifically for the purpose of comparing the calculated values for the correction factors.

For a given sample, the amount of scattered light which reaches the phototube depends on the beam collimation and on the relative position of source, sample, and phototube. For this reason, the results depend on the particular spectrophotometer and circular dichroism instrument used. The circular dichroism instrument used in this laboratory is the Cary 60 with the CD attachment containing the phototube in the sample compartment. This allows placement of the sample close to the phototube such that a greater amount of the forward scattered light is captured. By sonication, the extent of the distortion can be decreased, as seen in Figs. 3 and 4. A membrane sample prior to sonication has less absorbance at 190 nm (due to greater absorption flattening) but a higher measured absorbance at 240 nm (due to more light scattering) than after sonication such that the curves commonly cross over between these two wavelengths. The crossover, or isosbestic point, between the two absorption curves is a sensitive measure of the relative amounts of absorption flattening and light scattering. For curves a and c in Fig. 7, the absorption crossover occurred at 210 nm. When using the Cary 14 and positioning the cell closest to the phototube, the crossover of curves a and c also occurred near 210 nm (see Fig. 8). With the sample closest to the source, the crossover in the Cary 14 is at shorter wavelengths.

As seen in Table II, the calculated correction factors are almost identical when using the data in Figs. 7 and 8. Accordingly, it should be possible, in general, to verify the use of a particular spectrophotometer with a specified cell position as long as the CD instrument used has a means of obtaining absorbance, as, for example, when there is a dynode voltage meter on the CD photomultiplier.

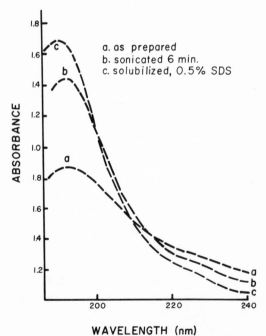

WAVELENGTH (nm)

Fig. 7. The simultaneously recorded absorption spectra obtained concurrently with those in Fig. 4 on the Cary 60 spectropolarimeter. It is to be compared with spectra obtained immediately afterward on the Cary 14 (Fig. 8).

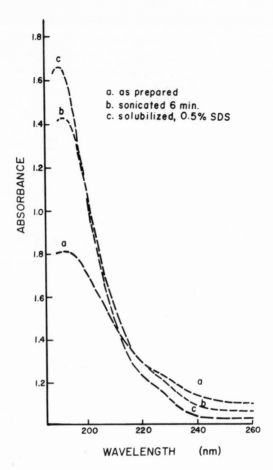

Fig. 8. Absorption spectra obtained with the Cary 14 spectrophotometer. The data were collected on the same samples as those in Fig. 4 and 7, immediately after the CD experiment. The protein concentration and the path length are the same as in Fig. 4. Note the correspondence in the absorbance data from the Cary 14 and the Cary 60. The cell is placed in the sample compartment but as close to the phototube compartment as possible.

a. as prepared
b. sonicated 6 min.
c. solubilized, 0.5% SDS

6. ANALYSIS OF CORRECTED MEMBRANE DATA

Efforts to interpret optical rotation data on membranes have followed the approach applied to soluble proteins, that is, a resolving of the data into contributions due to α-helical, β-type, and disordered or so-called random-coil patterns. This approach is faced with all of the limitations that have been repeatedly noted by many authors. The situation with corrected data on membrane preparations is obviously more tenuous than that with data on soluble homogeneous proteins. The reasons for this will be dealt with in some detail below. Briefly, the approach does not yet take into consideration specialized conformations involved in specialized membrane functions, it

Table II. Calculation of Correction Factors for Fig. 4

Calculation[a] of quantities	Curve a		Curve b	
	Cary 60 (OD-CD)[b]	Cary 14[b]	Cary 60 (OD-CD)[b]	Cary 14[b]
Q_A^{192}	0.49	0.50	0.86	0.89
A_P^{192}	1.84	1.76	0.20	0.14
A_P^{224}	0.18	0.20	0.02	0.016
Q_A^{224}	0.87	0.86	0.97	0.97
$(Q_A^{224})^2$	0.76	0.74	0.94	0.94
A_{susp}^{224}	0.27	0.26	0.23	0.25
A_F^{224}	0.14	0.15	0.16	0.17
$^oA_S^{224}$	0.13	0.11	0.07	0.08
Q_σ^{224}	0.75	0.77	0.83	0.82
Q_E^{224c}	0.57	0.57	0.78	0.77

[a] The calculations used the vesicle values in Table I.
[b] Use of the Cary 60 indicates the simultaneous measurement of absorbance and ellipticity using the same phototube, whereas the use of the Cary 14 is for absorbance data alone with the cell positioned in the sample compartment at the extreme phototube side of the cell holder.
[c] A most satisfying feature when applying this correction factor to the two membrane curves in Fig. 4 is that both curves correct to closely the same values at 224 nm. Also, of course, this demonstrates that the Cary 14 absorption data obtained with the proper cell position may be used to calculate the required quantities for correcting biomembrane data. Other CD and absorption instruments may be utilized in a similar manner.

does not account for end effects and for interactions between a given conformation and details of its environment, and it does not allow for contributions of chromophores other than the peptide. The treatment also assumes a static model.

6.1. Specialized Structures Proposed for Specialized Functions

Components within a biological membrane have very specialized functions, many of which are not demonstrated by soluble proteins and polypeptides. This consideration calls for caution when interpreting data on membranes in terms, for example, of α-helix, β-structures, and random coils. There is now more explicit evidence for exercising caution. One of the major functions of biological membranes is to provide selective permeation and thereby to control the intracellular environment. In addition, selective movement of ions is fundamental to mechanisms of cellular excitability and energy transformation. Two mechanisms of membrane permeation by cations, i.e., utilizing carriers and channels, have been shown to involve polypeptides and depsipeptides, and specialized conformations are involved in both mechanisms.

The proposed conformation of valinomycin when it resides at the membrane–aqueous environment interface is a series of three β-turns (see Fig. 9). On interacting with potassium ion, acyl oxygens of the β-turns close on the cation, like a three-pronged bear trap. The result is a bare cation held in the polar core of six octahedrally arranged acyl oxygens. Each acyl oxygen derives from the end of a β-turn (see Fig. 10). While the conformational feature we refer to as a β-turn can be expected to be prominent when polypeptide and protein bind cations, it is also observed in solubilized proteins. The crystal structure of cytochrome c, for example, contains many β-turns (Dickerson et al., 1971). This conformational feature has not been explicitly considered, prominent though it may be, when attempting to resolve circular dichroism and absorption data into various percent conformations.

Recently, a new set of unique helices have been proposed which may directly function as channels for selective ion permeation through membranes. In all previously described helices—for example, the 3_{10}-helix, the α-helix,

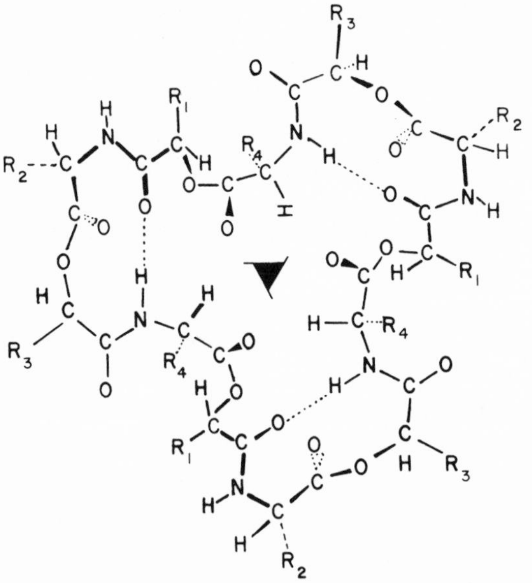

Fig. 9. Proposed conformation of one of several closely equivalent conformations of valinomycin in DMSO. The conformations are such that the back side is lipophilic, containing the aliphatic isopropyl and methyl side-chains, and the front side is hydrophilic due to the acyl oxygens of the ester moieties. The three hydrogen bonds define three β-turns. From Urry and Ohnishi (1970).

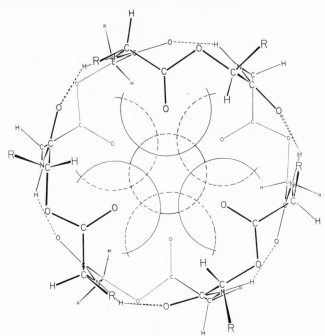

Fig. 10. Calculated structure of the lowest energy conformation of a valinomycin–cation complex in which the R groups were taken as methyls. The constraints imposed in calculating the structure were the standard dimension of the residues involved, C_3 symmetry, and an $O_i \ldots H_{i+3}$ hydrogen bond distance of 1.80 Å. The center circle corresponds to the K^+ drawn to scale. It can be seen that the size of the polar core containing the ion is controlled by the tilt of the ester moiety. The conformational energy changes brought about by tilting the ester moiety and changing the size of the polar core indicate a selectivity for K^+ over Na^+. From Mayers and Urry (1972).

the π-helix, the γ-helix—the peptide C-O moieties point toward the same end of the helical structure. In the newly described set of helices, the C → O bond axes alternately point parallel and antiparallel to the helix axes (Urry, 1971; Urry et al., 1971a). These structures are termed "β-helices" (Urry, 1972b, 1973b), as the hydrogen bonding pattern between turns of the helix is the same as between chains of a parallel β-pleated-sheet structure. A previous name, "$\pi_{L,D}$-helices" (used because the initial member of the series had the same number of residues per turn as the π-helix), has been changed for the more descriptive term "β-helix." The left-handed $\beta_{3,3}^6$-helix proposed for gramicidin A is given in Fig. 11. In terms of this helix, the mechanism whereby potassium ions move along the channel is by an ion-induced relaxation of the helical coordinates. The movement is one in which the

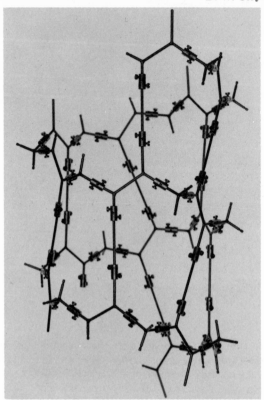

Fig. 11. Wire model of the $\beta_{3,3}^6$-helix showing the left-handed helical backbone and the intraturn hydrogen bonding pattern, which is the same as in the parallel β-pleated-sheet conformations. It is also apparent that the C-O moieties with bond axes directed toward the carboxyl end are parallel to the helix axis. The result is a net dipole moment. From Urry (1973b).

peptides librate such that the acyl oxygens move inward toward the center of the channel to interact with the bare ion, but they do so without breaking the intraturn hydrogen bonds. Again, should this molecule represent a model for natural membrane channels (Goodall and Sachs, 1972), the contribution of this conformation to the CD patterns of membranes requires consideration. Furthermore, additional new structures called "β-spirals," which are a regularly related series of β-turns that can readily interconvert to β-helices (see Fig. 12), may also be present in biomembranes (Urry, 1972b). The fundamental point of the above discussion is the realistic expectation, already partly realized, of specialized structures involved in the special functions essential to viable biological membranes.

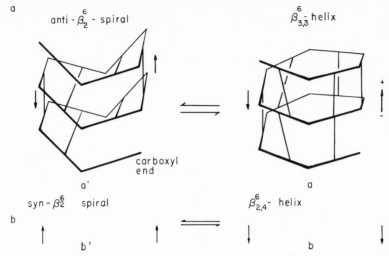

Fig. 12. Representation of the conversion from a spiral to a helical conformation noting the effect on the dipole moments arising from the peptide moieties. (a) In the anti-β_2^6-spiral, the end peptide moieties point in opposite direction such that their dipole moments cancel out, as do those of the cross-β peptides in which the peptide plane is nearly perpendicular to the spiral axis. However, on conversion to the $\beta_{3,3}^6$-helix, the peptides point in opposite directions, but in one direction the peptide C-O bond vectors are parallel to the helix axis, whereas the C-O bond vectors of the peptide pointing in the opposite direction are at an appreciable angle to the helix axis. This results in a net dipole moment along the helix axis (see Fig. 11). (b) The syn-β_2^6-spiral has both end-peptide C-O moieties pointing in the same direction, resulting in a large net dipole moment for this conformation. On conversion to the $\beta_{2,4}^6$-helix, four peptide C-O moieties point in the opposite direction, while the two that originated from the end peptides of the spiral structure maintain the direction but skew slightly with respect to the helix axis. The result is that both conformations have large net dipole moments but in opposite directions. From Urry (1971).

6.2. Domains

In a general approach to the analysis of data on a complex membrane system, the term "domain" is utilized. The membrane is considered to be a sum of domains. A domain is defined as any reasonable and convenient subdivision of the membrane for the analysis of immediate concern. For example, an α-helical segment could be considered a domain. The macrocopically observed molar quantity, $\langle Z \rangle$, for a membrane system is the sum of the molar quantities for each domain (Z_i) times the mole fraction (x_i) that the domain represents; i.e.,

$$\langle Z \rangle = \sum_i x_i Z_i \tag{42}$$

The usual calculation of percent α-helix is based on this form of expression; that is, when one assumes two states, one α-helical and the other disordered, the observed mean residue ellipticity, $[\theta]_{obs}^{\lambda}$, at a given wavelength λ, becomes

$$[\theta]_{obs}^{\lambda} = x_H [\theta]_H^{\lambda} + x_D [\theta]_D^{\lambda} \tag{43}$$

where $[\theta]_H^{\lambda}$ and $[\theta]_D^{\lambda}$ are the mean residue ellipticities from model α-helical and disordered proteins and polypeptides. Since $x_H + x_D = 1$, one readily solves for the mole fraction of helix:

$$x_H^{\lambda} = \frac{[\theta]_{obs}^{\lambda} - [\theta]_D^{\lambda}}{[\theta]_H^{\lambda} - [\theta]_D^{\lambda}} \tag{44}$$

The equation is written for a given wavelength. If the assumption of only two noninteracting states with adequately determined reference or model states were correct, the statement should be correct for each wavelength. This simple test of adequacy quickly shows the assumptions to be limiting. What one has, of course, is a greater number of domains and interaction between domains. The general expression for calculating the mole fraction of a given domain is

$$x_i = \frac{[\theta]_{obs}^{\lambda} - \sum_{j \neq i} (1 + \sum_{k \neq i,j} x_k) [\theta]_j^{\lambda}}{[\theta]_i^{\lambda} - \sum_{j \neq i} [\theta]_j^{\lambda}} \tag{45}$$

An analogous expression may also be written involving the molar extinction coefficient:

$$x_i = \frac{\varepsilon_{obs}^{\lambda} - \sum_{j \neq i} (1 + \sum_{k \neq i,j} x_k) \varepsilon_j^{\lambda}}{\varepsilon_i^{\lambda} - \sum_{j \neq i} \varepsilon_j^{\lambda}} \tag{46}$$

As the membrane is a dynamic structure, it should be recognized that, at a given temperature, there may exist an equilibrium between two different conformations of a given domain, as, for example, the interconversion between β-spiral and β-helical conformations (see Fig. 12). In order to account for this effect, the probability P_{iq} that the ith domain is in the q conformation is considered by taking appropriate cognizance of the energy of the qth state for the ith domain (ε_{iq}) relative to the sum of all the allowable energy states for the domain:

$$P_{iq} = \frac{e^{-\varepsilon_{iq}/kT}}{\sum\limits_{q} e^{-\varepsilon_{iq}/kT}} \tag{47}$$

With this consideration, the average property becomes

$$\langle Z \rangle = \sum_{i} X_i \sum_{q} Z_{iq} P_{iq} \tag{48}$$

where

$$\sum_{q} Z_{iq} P_{iq} = Z_i \tag{49}$$

Basic quantities when interpreting absorption and circular dichroism data and when relating to theoretical work are the dipole strength (D_i) and the rotational strength (R_i). The quantities are approximated from the spectra with the following equations (Moscowitz, 1960):

$$D_i = 1.63 \times 10^{-38} \frac{\varepsilon_i^0 \, \varDelta_i}{\lambda_i} \tag{50}$$

and

$$R_i = 1.23 \times 10^{-42} \frac{[\theta_i^0] \, \varDelta_i}{\lambda_i} \tag{51}$$

where ε_i^0 and $[\theta_i^0]$ are the molar extinction and molar ellipticity, respectively, at the maximum of the ith band occurring at wavelength λ_i and $\varDelta_i$ is the half bandwidth at magnitudes of ε_i^0/e and $[\theta_i^0]/e$.

A quantity which has been utilized in the past to compare bands in the absorption and circular dichroism, often for the purpose of aiding in assignment of electronic transitions, is called the "anisotropy" (g_i) of the ith band. It is defined as the ratio of the rotational strength to the dipole strength,

$$g_i = \frac{R_i}{D_i} \tag{52}$$

which, utilizing the above equations, becomes

$$g_i = 0.756 \times 10^{-4} \frac{[\theta_i^0]}{\varepsilon_i^0} \tag{53}$$

The spectra of biomembranes are the contribution of several electronic transitions; for example, the absorbance near 222 nm in membranes is due to an $n–\pi^*$ transition of peptides and to contributions from aromatic side–

chains and lipid ester moieties. For this reason, it is perhaps more reasonable to refer to critical wavelength anisotropies (g_i^λ) rather than band anisotropies. The point of interest in a critical wavelength anisotropy is in its potential in providing delineation of complex mixtures of conformations where ellipticity alone or extinction coefficients alone would be less sensitive. This is because while the magnitude of molar ellipticity at 190 nm may be increasing as α-helix is formed, the molar extinction coefficient at 190 nm is decreasing. Conversely, when going from α-helix to a β-pleated-sheet conformation, the molar ellipticity at 190 nm decreases whereas the molar extinction coefficient increases. Accordingly, the ratio would provide an experimental quantity more sensitive to changes in conformation. As with the individual quantities, the mole fraction of a particular conformation can be calculated from suitable reference state values; i.e.,

$$x_i = \frac{g_{obs}^\lambda - \sum_{j \neq i}(1 + \sum_{k \neq i,j} x_k)\, g_j^\lambda}{g_i^\lambda - \sum_{j \neq i} g_j^\lambda} \tag{54}$$

Application of this approach to complex biomembranes with the purpose of identifying changes in polypeptide backbone conformation necessitates removal of lipid and aromatic side-chain contributions to the measured absorption spectra. The common assumption when discussing the optical activity of biomembranes is to neglect the contribution of lipid and aromatic groups. When such an assumption is justified, one can further assume, in the various interactions which lipid chromophores and amino acid aromatic side-chains experience in a biomembrane, that on the average the hyper- and hypochromism of a given band average out and that one can use the values obtained on the isolated chromophores. Following this approach for red blood cell ghosts, it is apparent that aromatic side-chains of amino acids constitute the largest correction to the absorption spectra, with lipid contributing only about 20 % of the absorbance (Long et al., unpublished results).

6.3. Empirical Correlations

6.3.1. Comparison of Membrane Systems

When comparing corrected ellipticity data on a series of biomembranes —specifically red blood cell ghosts, heavy beef-heart mitochondria, plasma membranes, sarcotubular vesicles, and axonal membranes—there is observed a trend from higher to lower ellipticity magnitudes at 224 nm (Urry, 1972a; Urry et al., 1971b). In noting the series, one also recognizes a trend from more

enzymatic membrane systems to those which are specialized for ion movement. While in the given order the magnitude of ellipticity decreases, there
is, in general, an inverse situation in absorbance. The absorbance per milligram of protein in the membranes increases, with the exception of the
sarcotubular vesicles. Once these membrane data have been adequately
corrected for lipid and aromatic absorbances, a more informative detailed
comparison can be made utilizing the values for the critical wavelength
anisotropies (Masotti *et al.*, unpublished results). From the gross data, it
would already appear that sarcotubular vesicles are an interesting system,
as they seem not to be readily characterized in terms of the usual considerations of α-helix, β-structures, and random-coil conformations.

6.3.2. Substrate and Cofactor Effects on Membrane Structure

Efforts to observe changes in the CD pattern of biomembranes as a result
of substrate or cofactor interactions have generally led to little more than the
observation of aggregation effects resulting from associated divalent cations

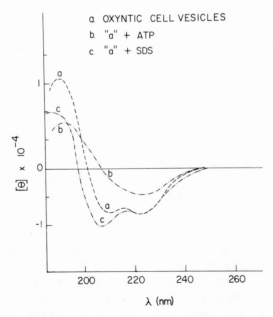

Fig. 13. The CD spectrum of vesicles derived from oxyntic cells, of vesicles plus Mg ATP
(1 mM) (uncorrected), and of vesicles plus sodium dodecylsulfate (SDS). From Masotti
et al. (1972).

Mg^{2+} and Ca^{2+}. While the aggregation-mediated effects of divalent cations may contribute to our knowledge of membrane structure, they are not the specific effects that many have sought. Two membrane systems, however, have demonstrated specific substrate or cofactor responses. One is a membranous preparation from the oxyntic cells of the gastric mucosa and the other is a preparation of synaptosomal ghosts. The synaptosomal ghost preparation responds to nicotine plus Ca^{2+} at divalent cation levels well below those which cause dramatic changes in aggregation (Saccomani and Urry, unpublished results). The oxyntic cell preparation responds to Mg ATP, but not to a similar concentration of Mg AMP (Masotti et al., 1972). The oxyntic cell data are given in Fig. 13. Whether Mg ATP is itself causing a change in protein conformation or a change in the state of aggregation can only be decided once adequate confidence is achieved in the applied corrections. The calculated values in Fig. 4 at 224 nm indicate that there is a basis for confidence in the values for corrected biomembrane spectra.

6.3.3. Solubilization and Reconstitution

As studies progress on the solubilization of membrane components in enzymatically or otherwise active states and as efforts are made to achieve reconstitution, circular dichroism and absorption spectra can serve as criteria for determining whether such processes have occurred with or without conformational change. At this stage, there is little to report on such efforts, but one or two relevant points may be made. In spectroscopic studies on the conformation of gramicidin A (Urry et al., 1972; Glickson et al., 1972), it was found that in trifluoroethanol and dimethylsulfoxide a good argument existed for the β-helical conformation. Since the conformation of gramicidin A is variable with solvent system, it becomes of great interest to assess the conformation of gramicidin A when incorporated into lipid vesicles. Incorporation can be achieved by sonication of lecithin micelles in the presence of gramicidin A. Analysis of the corrected spectra will provide the desired information.

The second point concerns the $(Na^+ + K^+)$ ATPase from brain. In efforts to correct this spectra, sodium dodecylsulfate was used to solubilize the membrane in a manner sufficient to satisfy the criteria for a pseudo reference state. It was later found that the Triton X-100 solubilized preparation which was enzymatically active exhibited a CD pattern almost identical to that which was previously judged an adequate pseudo reference state. This result adds to the credibility of the corrections and the use of circular dichroism to assess solubilization and reconstitution efforts.

ACKNOWLEDGMENTS

This study was aided by a grant from the American Medical Association Education and Research Foundation and by financial support from the Mental Health Board of Alabama. One of us (M.M.L.) was supported by the National Institutes of Health, Postdoctoral Fellowship AM52457-01.

7. REFERENCES

Bayliss, N. S., and McRae, E. G., 1954, Solvent effects in organic spectra. Dipole forces and the Franck–Condon principle, *J. Phys. Chem.* **58**:1002.

Dickerson, R. E., Takano, T., Eisenberg, D., Kallai, O. B., Samson, L., Cooper, A., and Margoliash, E., 1971, Ferricytochrome *c*. I. General features of the horse and bonito proteins at 2.8 Å resolution, *J. Biol. Chem.* **246**:1511.

Duysens, L. N. M., 1956, The flattening of the absorption spectrum of suspensions, as compared to that of solutions, *Biochim. Biophys. Acta* **19**:1.

Glaser, M., and Singer, S. J., 1971, Circular dichroism and the conformations of membrane proteins studied with red blood cell membranes, *Biochemistry* **10**:1780.

Glickson, J. D., Mayers, D. F., Settine, J. M., and Urry, D. W., 1972, Spectroscopic studies on the conformation of gramicidin A: PMR assignments, coupling constants, and H-D exchange, *Biochemistry* **11**:477.

Goodall, M., and Sachs, G., 1972, Excitable tissue—Extraction of K^+ selective channels, *Nature* **237**:252.

Gordon, D. J., 1972, Mie scattering by optically active particles, *Biochemistry* **11**:413.

Gordon, D. J., and Holzwarth, G., 1971, Artifacts in the measured optical activity of membrane suspensions, *Arch. Biochem. Biophys.* **142**:481.

Gordon, A. F., Wallach, D. F. H., and Strauss, J. H., 1969, Optical activity of plasma membranes and its modification by lysolecithin, phospholipase A, and phospholipase C, *Biochim. Biophys. Acta* **183**:405.

Grosjean, M., and Tari, M., 1964, Progrès dans la mesure du dichroisme circulaire optique, jusqu'à 1850 Å, *Compt. Rend. Acad. Sci.* **258**:2034.

Holzwarth, G., and Doty, P., 1965, The ultraviolet circular dichroism of polypeptides, *J. Am. Chem. Soc.* **87**:218.

Iizuka, E., and Yang, J. T., 1966, Optical rotatory dispersion and circular dichroism of the beta-form of silk fibroin in solution, *Proc. Natl. Acad. Sci. (U.S.)* **55**:1175.

Kasha, M., 1963, Energy transfer mechanisms and the molecular exciton model for molecular aggregates, *Radiation Res.* **20**:55.

Litman, B. J., 1972, Effect of light scattering on the circular dichroism of biological membranes, *Biochemistry* **11**:3243.

Long, M. M., Masotti, L., Sachs, G., and Urry, D. W., 1973, *J. Supramolecular Structure*, in press.

Long, M. M., Spitzer, H. L., and Urry, D. W., unpublished results.

Masotti, L., Long, M. M., Sachs, G., and Urry, D. W., 1972, The effect of ATP on the circular dichroism spectrum of membrane fractions from oxyntic cells, *Biochim. Biophys. Acta* **255**:420.

Masotti, L., Urry, D. W., and Llínas, I., unpublished results.

Mayers, D. F., and Urry, D. W., 1972, The valinomycin–cation complex: Conformational energy aspects and ion selectivity, *J. Am. Chem. Soc.* **94**:77.

Mie, G., 1908, Contribution to the optics of tubid media, especially colloidal metal solutions, *Ann. Phys.* **25**:377.

Moscowitz, A., 1960, Theory and analysis of rotatory dispersion curves, *in* "Optical Rotatory Dispersion" (C. Djerassi, ed.), p. 150, McGraw-Hill, New York.

Ottaway, C. A., and Wetlaufer, D. B., 1970, Light-scattering contributions to the circular dichroism of particulate systems, *Arch. Biochem. Biophys.* **139**:257.

Quadrifoglio, F., and Urry, D. W., 1968a, Ultraviolet rotatory properties of polypeptides in solution. I. Helical poly-L-alanine, *J. Am. Chem. Soc.* **90**:2755.

Quadrifoglio, F., and Urry, D. W., 1968b, Ultraviolet rotatory properties of polypeptides in solution. II. Poly-L-serine, *J. Am. Chem. Soc.* **90**:2760.

Rhodes, W., 1961, The hypochromism and other spectral properties of helical poly-nucleotides, *J. Am. Chem. Soc.* **83**: 3609.

Saccomani, G., and Urry, D. W., unpublished results.

Sarkar, P. K., and Doty, P., 1966, The optical rotatory properties of the β-configuration in polypeptides and proteins, *Proc. Natl. Acad. Sci. (U.S.)* **55**:981.

Skou, J. C., 1965, Enzymatic basis for active transport of Na^+ and K^+ across cell membranes, *Physiol. Rev.* **45**:596.

Tinoco, I., 1962, Theoretical aspects of optical activity. II. Polymers, *Advan. Chem. Phys.* **4**:113.

Tomimatsu, Y., Vitello, L., and Gaffield, W., 1966, Effect of aggregation on the optical rotatory dispersion of poly(α, L-glutamic acid), *Biopolymers* **4**:653.

Urry, D. W., 1965, Protein–heme interactions in heme proteins: Cytochrome C, *Proc. Natl. Acad. Sci (U.S.)* **54**:640.

Urry, D. W., 1967a, The heme chromophore in the ultraviolet, *J. Biol. Chem.* **242**: 4441.

Urry, D. W., 1967b, Model systems for interacting heme moieties. I. The heme undecapeptide of cytochrome C, *J. Am. Chem. Soc.* **89**: 4190.

Urry, D. W., 1968, Optical rotation, *Ann. Rev. Phys. Chem.* **19**:477.

Urry, D. W., 1970, Optical rotation and biomolecular conformation, *in* "Spectroscopic Approaches to Biomolecular Conformation" (D. W. Urry, ed.), p. 33, American Medical Association Press, Chicago.

Urry, D. W., 1971, The gramicidin A transmembrane channel: A proposed $\pi_{(L,D)}$ helix, *Proc. Natl. Acad. Sci. (U.S.)* **68**:672.

Urry, D. W., 1972a, Protein conformation in biomembranes: Optical rotation and absorption of membrane suspensions, *Biochim. Biophys. Acta* **265**:115.

Urry, D. W., 1972b, A molecular theory of ion conducting channels: A field dependent transition between conducting and nonconducting conformations, *Proc. Natl. Acad. Sci. (U.S.)* **68**:1907.

Urry, D. W., 1973a, Corrections for optical rotation data of biomembranes, *Meth. Enzymol.*, in press.

Urry, D. W., 1973b, *The Jerusalem Symposia on Quantum Chemistry and Biochemistry*, V. The Israel Academy of Science and Humanities, Jerusalem, 1973, p. 723.

Urry, D. W., and Ji, T. H., 1968, Distortions in circular dichroism of particulate (or membranous) systems, *Arch. Biochem. Biophys.* **128**:802.

Urry, D. W., and Krivacic, J., 1970, Differential scatter of left and right circularly polarized light by optically active particulate systems, *Proc. Natl. Acad. Sci. (U.S.)* **65**:845.

Urry, D. W., and Ohnishi, M., 1970, Nuclear magnetic resonance and the conformation of cyclic polypeptide antibiotics, *in* "Spectroscopic Approaches to Biomolecular Conformation" (D. W. Urry, ed.), p. 263, American Medical Association Press, Chicago.

Urry, D. W., and Pettegrew, J. W., 1967, Model systems for interacting heme moieties. II. The ferriheme octapeptide of cytochrome C, *J. Am. Chem. Soc.* **89**: 5276.

Urry, D. W., and van Gelder, B. F., 1968, Circular dichroism of cytochrome oxidase and the CO and CN derivatives, *in* "Structure and Function of Cytochromes" (K. Okunuki, M. D. Kamen, and I. Suzuki, eds.), p. 210, University Park Press, Baltimore.

Urry, D. W., Wainio, W. W., and Grebner, D., 1967, Evidence for conformational differences between oxidized and reduced cytochrome oxidase, *Biochem. Biophys. Res. Commun.* **27**:625.

Urry, D. W., Ruiter, A. L., Starcher, B. C., and Hinners, T. A., 1968, "Antimicrobial Agents and Chemotherapy," p. 87. (G. L. Hobby ed.) American Society for Microbiology, Bethesda, Maryland.

Urry, D. W., Hinners, T. A., and Masotti, L., 1970a, Calculation of distorted circular dichroism curves for poly-L-glutamic acid suspensions, *Arch. Biochem. Biophys.* **137**:214.

Urry, D. W., Masotti, L., Krivacic, J., 1970b, Improved ellipticity data for several biological membranes, *Biochem. Biophys. Res. Commun.* **41**: 521.

Urry, D. W., Goodall, M. C., Glickson, J. D., and Mayers, D. F., 1971a, The gramicidin A transmembrane channel: Characteristics of head to head dimerized $\pi_{(L,D)}$ helices, *Proc. Natl. Acad. Sci. (U.S.)* **68**:1907.

Urry, D. W., Masotti, L., and Krivacic, J. R., 1971b, Circular dichroism of biological membranes. I. Mitochondria and red blood cell ghosts, *Biochim. Biophys. Acta* **241**:600.

Urry, D. W., Glickson, J. D., Mayers, D. F., and Haider, J., 1972, Spectroscopic studies on the conformation of gramicidin A: Evidence for a new helical conformation, *Biochemistry* **11**:677.

Yanari, S., and Borey, F. A., 1960, Interpretation of the ultraviolet spectral changes in proteins, *J. Biol. Chem.* **235**: 2181.

Chapter 4

Isolation and Serological Evaluation of HL-A Antigens Solubilized from Cultured Human Lymphoid Cells

R. A. REISFELD, S. FERRONE, AND M. A. PELLEGRINO

Department of Experimental Pathology
Scripps Clinic and Research Foundation
La Jolla, California

1. INTRODUCTION

Histocompatibility (H) antigens are genetically segregating cell surface markers which elicit an immune response by the host following the transplant of foreign tissues. Allografting thus furnishes an excellent tool for distinguishing these antigenic cell surface markers, as it causes the host to recognize a set of cellular antigens which distinguish him from the graft donor.

These H-antigenic specificities are under the control of genetic loci and are extremely polymorphic in the outbred human population, constituting an allotypic system, i.e., a group of antigenic specificities shared by some but not all members of a species. There is in most species a single major strong histocompatibility locus controlling the rapid rejection of allografts. These strong loci code for a mosaic of serological specificities. In man, this locus is called "HL-A" (human leukocyte, locus A) and is considered to be organized into two segregant series: one, comprised of at least six well-recognized specificities (HL-A1,2,3,9,10,11), being designated LA and the other, comprised of at least five specificities (HL-A5,7,8,12,13), named "FOUR" (Dausset *et al.*, 1970; Kissmeyer-Nielsen and Thorsby, 1970).

Additional antigenic determinants controlled by these loci are still being investigated.

The concept of two segregant series is consistent with data from population analyses and recombinant studies in families. The interest in this antigenic system is partly due to observations implying that these individuality markers are useful when selecting organ donors from among a transplant recipient's siblings. Among siblings, there is a definite correlation between HL-A phenotype and tissue graft survival; a close match of the H-antigenic determinants of donor and host provides a better chance of transplant survival (for review see Kissmeyer–Nielsen and Thorsby, 1970).

The actual physiological function of these cell surface located H-antigens is most likely independent of the artificial situation of transplantation immunity. Due to their strategic location and their relatively strong immunogenicity, they probably exert an important role in cell economy as receptor sites in immunosurveillance, growth regulation, permeability, and cell contact phenomena. Most prominent among the various theories is the concept of immunosurveillance, which proposes that H-antigens act as surface receptors permitting the immune system to distinguish autologous from foreign cells (Thomas, 1959; Burnet, 1970). It has also been suggested that messenger lymphocytes bearing immunoglobulin determinants structurally complementary to H-antigens might be instrumental in deciphering the mosaics of cell surface receptors in the body (Mason and Warner, 1970). Aside from their potentially fundamental role in cell physiology, the genetically determined H-antigens are most likely an integral part of the cytoarchitecture *per se*. Through their inadvertent participation in allograft immunity, they offer one of the most specific systems for the study of cell membranes. Thus a thorough elucidation of the serological profiles and chemical characteristics of H-antigens should contribute to a better understanding of cell membrane structure and function.

2. EXTRACTION OF SOLUBLE HISTOCOMPATIBILITY ANTIGENS

2.1. Cell Surface Location

There is a large body of evidence which indicates that the alloantigenic determinants controlled by H-loci are arrayed on cell surface membranes. Early studies with agglutinating alloantibody had suggested that the major portion of the strong alloantigenic determinants were associated with the cell surface (Amos, 1953; Gorer and Mikulska, 1954). A number of observations confirmed this location, in particular studies with (1) fluorescent antibody (Möller, 1961; Cerottini and Brunner, 1967; Gervais, 1968), (2) antibody absorption prior to and after cell rupture (Haughton, 1964), and (3) purified membrane fractions (Ozer and Hoelzl-Wallach, 1967). To get an idea of the cell surface location of alloantigens, Boyse et al. (1968) employed antibody blocking reagents recognizing five murine alloantigenic systems (H-2 θ, LY-A, Ly-B, and TL). The studies suggested that these specificities comprise a single cluster or basic segment of the repetitive membrane structure. The proximity of allotypic specificities was confirmed for H-2 with ferritin-tagged antibodies (Davis and Silverman, 1968) and for HL-A with an indirect ferritin-labeling method (Kourilsky et al., 1971) and with an indirect fluorescence assay (Ferrone et al., 1972a). When Boyse et al. (1968) applied their antibody blocking test, they found that H-2 antigens controlled at either the H-2D or H-2K locus did overlap on lymphocytes, in contrast to thmrocytes where such determinants were found distant from each other in separate clusters. The much greater concentration of H-2 determinants on lymphocytes apparently results in a more confluent pattern. However, when employing ferritin-labeled alloantibodies, separate clusters of H-2^d, H-2^k, and H-2^b antigens were detected on lymphocytes (Davis and Silverman, 1968; Aoki et al., 1970). Similarly, by indirect tests, HL-A antigens were detected as patches on the membrane of peripheral lymphocytes (Kourilsky et al., 1971) and of cultured lymphoid cells (Ferrone et al., 1972a; Pellegrino et al., 1972b). However, a continuous distribution of HL-A antigens was stipulated on the basis of data from indirect sandwich-type labeling techniques (Willingham et al., 1971). Furthermore, Davis et al. (1971) demonstrated that the distribution of H-2 antigens appeared continuous when examined by a direct ferritin-conjugated antibody technique but discontinuous when an indirect technique was utilized. Some possible causes of this phenomenon have been advanced (Davis et al., 1971; Davis, 1972): (1) γ-globulins which remain nonconjugated with ferritin may block part of the cell surface (Aoki et al., 1969); (2) certain membrane regions are inaccessible to antibody due to "folding" during the labeling process; (3) H-antigens as well as other membrane components lack any fixed position on the cell membrane and continually are in a fluid rearrangement (Frye and Edidin, 1970; Singer and Nicolson, 1972). In addition, it is possible that the patchlike distribution represents an artifact of the ferritin or fluorescent procedure. Consequently, it seems worthwhile to exercise some caution when using this particular finding as the sole basis for new hypotheses concerning the expression of H-antigens on cell surfaces.

2.2. Criteria of Solubility

The concept of solubilization of H-antigens from cell membranes is based largely on the assumptions that (1) antigenic determinants can be isolated independent of membrane structures, (2) there are no immunologically significant associations between solubilized antigens and other membrane components, and (3) the solubilized material represents the antigenic moiety with its determinants in a form more or less similar to that originally present on the cell membrane.

There are some differences of opinion as to what actually constitutes a soluble antigen. Some investigators consider an antigen to be in soluble form when it fails to sediment upon ultracentrifugation at $100,000 \times g$, i.e., if it is free of any visible cell membrane fragments. This concept is questionable, especially since Rapaport *et al.* (1965) found membrane fragments in antigens which did not sediment at $100,000 \times g$ by electron microscopic examination of the $200,000 \times g$ sediment of these "soluble antigens." Other investigators defined "as an acceptable demonstration of solubility that the antigenic material should be able to pass through a Sephadex gel filtration column and not be eluted with the void volume" (Blandamer *et al.*, 1969). The behavior of antigenic material on Sephadex columns is a questionable criterion of antigen solubility because these materials aggregate easily and thus can appear in the void volume even when soluble.

The solubility of a substance is best defined by its characteristic solubility in any given solvent. Since most H-antigen preparations isolated thus far are complex mixtures of proteins and conjugated proteins subject to intricate protein–protein interactions, classical solubility characteristics such as those used to define protein purity, e.g., solubility curves and light-scattering techniques, are hardly applicable. Consequently, at present it seems reasonable to define the solubility of H-antigens in practical terms: by the lack of membrane fragments detected by ultrastructural methods and by the ability of such methods as ion-exchange chromatography and electrophoresis to resolve antigens *without* serious loss of biological activity due to insolubility at varying protein concentrations.

There have been a number of attempts to solubilize H-antigens from cell membranes by such diverse techniques as organic solvent and detergent treatment (for reviews, see Reisfeld and Kahan, 1970*a*, 1971), digestion with proteolytic enzymes (for reviews, see Reisfeld and Kahan, 1970*a*, 1972*b*; Mann and Fahey, 1971; Sanderson and Welsh, 1972*a*), application of sonic energy (for review, see Kahan and Reisfeld, 1971), treatment with complex salts (Mann, 1972), and extraction with simple salts (Reisfeld *et al.*, 1971*a*; Reisfeld and Pellegrino, 1972). Since the application of most of these methods has already been extensively described and reviewed, we will detail mainly a solubilization procedure for HL-A antigens developed in our laboratory which utilizes the extraction of cultured cells with 3 M KCl.

2.3. Antigen Source Material

Although inbred animal lines have been the key to understanding transplantation antigen systems in such species as mice and guinea pigs, they have not provided an adequate source for extraction of sufficient amounts of H-antigens to permit extensive chemical characterization.

In the case of human H-substances, the situation was even more difficult since one dealt with an outbred population exhibiting a tremendous degree of polymorphism as far as HL-A determinants are concerned. Until quite recently, there was no source available from which one could adequately extract large amounts of soluble HL-A antigens of uniform genetic constitution, as illustrated by the miniscule amounts of HL-A antigens extracted from individual human spleens (Kahan *et al.*, 1968; Sanderson, 1968; Davies, 1969). However, this difficulty was overcome by the discovery that lymphoid cells in long-term culture are almost ideal for the purpose (Mann *et al.*, 1968; Reisfeld *et al.*, 1970). Cultured human lymphoid cell lines derived from normal donors and from those with lymphoid malignancies were first perpetuated on a large scale by Moore *et al.* (1967).

Cell lines derived from a normal donor appear to be preferable to those obtained from a donor with lymphoid malignancy in view of (1) the hazard of transmitting malignant agents to normal patients during biological studies, (2) the possible relationship between tumor antigenic determinants and histocompatibility antigenic determinants, and (3) the need to have a living donor for long-term comparisons between peripheral lymphocytes, cultured cells, and antigen extracted from them.

However, a certain amount of caution has to be exercised when using cultured cells derived from normal donors for the extraction of HL-A antigens, especially if one plans to utilize them for injection into human volunteers, because recent work has demonstrated that all cultured human lymphoid cell lines possess at least a part of the Epstein–Barr virus genome;

it is thus felt that these lines carry in addition a foreign genetic load (zur Hausen *et al.*, 1972). It is difficult to assess the importance of this finding at the present time. However, it is of some interest that although E-B virus could readily be detected in the cultured cells used for antigen extraction, the KCl antigen extract was found to be devoid of it (P. Gerber, personal communication).

HL-A antigens can be readily solubilized from human spleen cells (Etheredge and Najarian, 1971) and from peripheral lymphocytes (E. E. Etheredge, personal communication; J. H. Pincus and M. A. Pellegrino, unpublished results). In addition, HL-A antigens were obtained in soluble form when human platelets were treated with 3 M KCl (Uhlenbruck *et al.*, 1973). The platelets, which are easily obtainable in large amounts from peripheral blood, provide potentially one of the best possibilities for obtaining adequate amounts of HL-A antigens from a single donor.

2.4. Extraction Techniques

The extraction methods which have thus far been applied to solubilize H-antigens from cells and tissues all have one serious limitation in common: they are nonspecific, i.e., they solubilize, aside from the antigenic moiety, vast amounts of contaminant materials. The reason for this is that the physicochemical properties and molecular nature of these antigens have not been elucidated thoroughly enough to permit the design of a specific extraction method.

Two basic strategies have been utilized to extract soluble H-antigens from cell surfaces: (1) covalent bond cleavage and (2) disruption of noncovalent intermolecular interactions.

2.4.1. Proteolytic Cleavage and Detergent Extraction

Successful antigen solubilization has been achieved with covalent cleavage by phospholipase A treatment of detergent-extracted membranous moieties and with proteolytic digestion of cellular membrane eluates (Kandutsch, 1960, 1963). Autolytic methods which relied on the action of cellular cathepsins also solubilized H-antigens, but in quite low yields of 1–2 % (Shimada and Nathenson, 1969) and with a relatively low degree of immunological specificity as determined by careful serological analyses (Sanderson, 1968). Concerted proteolysis with papain solubilized both human and murine H-antigens (Davies *et al.*, 1968; Sanderson and Batchelor, 1968; Mann *et al.*, 1968; Shimada and Nathenson, 1969). The antigenic moieties released by papain have been extensively studied; the yield of purified material is quite low: from 4000 mouse spleens estimated to contain approximately 40 mg of antigen, papain solubilization and subsequent electrophoretic purification retrieved three fractions containing a total of 1.43 mg H-2 antigen (Shimada and Nathenson, 1969). Although the specific activity of the antigenic material had been increased, approximately 98% of the total activity units present in the crude extract were lost. Mann et al. (1969) recovered 0.025 mg antigen following electrophoretic purification of papain-solubilized HL-A antigens from 3.8×10^9 cultured human lymphoid cells. Therefore, these investigators recommended that extraction methods based on covalent cleavage of peptide bonds by proteolytic enzymes be abandoned in favor of methods utilizing detergents such as sodium dodecylsulfate (Mann and Levy, 1971).

Aside from achieving relatively low yields of H-antigens, papain, a nonspecific proteolytic enzyme, not only cleaves the antigen from the membrane but may well cleave the antigen molecule itself. The choice of this enzyme as a first step in the solubilization of highly complex, membrane-associated H-antigens could preclude a good understanding of the structure of macromolecules on the cell surface, since it most likely destroys both subunit and primary structure.

On the other hand, methods which do not primarily depend on the cleavage of covalent bonds are apparently less destructive to H-antigenic structure and release relatively intact antigenic moieties. However, the application of detergents (Triton X-100, decyl- and dodecylsulfate, deoxycholate, sodium lauroyl sarcosinate, and Nonidet P-40) and the butanol extraction method yield essentially stabilized colloidal suspensions rather than truly soluble materials. These methods "solubilize" antigens in relatively low yields in an essentially water-insoluble state and with relatively poor specific antigenic activity (for review, see Reisfeld and Kahan, 1972a).

2.4.2. Low-Frequency Sound

Soluble materials with transplantation antigenic activity were first obtained by exposure of murine peripheral lymphocytes to low-frequency sonic energy, i.e., less than 16,000 cycles/sec (Kahan, 1965). Ultrasound (above 16,000 cycles/sec) denatured the determinants of H-antigens, probably because of the destructive effects of this energy on molecular conformation, including covalent bond cleavage and excessive local heating effects. The low-frequency sound is generated by a magnetostrictive oscillation (Raytheon model DF 101) with a plate voltage generating oscillations of a laminated rod sealed within the coil excitation field. The oscillations are transduced on a diaphragm generating sonic waves in the treated material. To achieve optimal yields of soluble antigen, one has to control carefully the conditions of sonication, including frequency, intensity, temperature, cell concentration, and exposure time. A brief exposure (3–5 min) of a cell suspension (25–50 $\times$ 10^6 cells/ml) to low-frequency sound (9–10 kc/sec, 15.5 W/cm^2) at 4°C liberates from 12 to 15% of the total antigenic activity of murine, guinea pig, canine, or human cells in soluble form (Reisfeld and Kahan, 1970a). The mechanism of sonic liberation, although not well understood, is thought to represent a depolymerization of noncovalent interactions between the antigen and insoluble cell membrane components (for review, see Kahan and Reisfeld, 1971).

2.4.3. Complex Salts

Recently, additional methods have been described for HL-A antigen solubilization which owe their effectiveness largely to noncovalent bond cleavage. These include (1) complex salts such as TIS, a tris salt of 2-hydroxy-3,5-diiodobenzoic acid (Mann and Levy, 1971), and (2) simple salts such as potassium chloride (Reisfeld and Kahan, 1970b; Reisfeld et al., 1971a).

TIS was reported to release 7–12% of the total HL-A antigenic activity detected on the surface of lymphocytes, compared to 15–20% release obtained from these cells by papain. While the material solubilized by papain appeared electrophoretically complex, the TIS extract seemed more uniform by this criterion (Mann, 1972). However, TIS is quite difficult to remove from proteins because of its diiodo groups, which at alkaline pH readily form covalent bonds with reactive amino acid moieties, e.g., sulfhydryl, phenolic hydroxyl, and amino groups. Additional chemical purification is impaired by this phenomenon. TIS effectively extracts HL-A antigens at a molarity of 0.1, while it destroys antigenic activity completely at 0.2 M, although solubilizing 80% of the cell membrane material (Mann, 1972).

2.4.4. Simple Salts

One of the more promising methods developed to obtain water-soluble histocompatibility antigens from the surface of cultured human lymphoid cells has involved the extraction of these cells with 3 M KCl (Reisfeld et al., 1971a). The choice of this method was based on earlier work on isolation of polypeptide hormones from human pituitary glands which showed that hypertonic salt extraction with 0.3 M KCl effectively extracted active hormone from these tissues (Reisfeld et al., 1962). As described below, this method can extract soluble HL-A antigen in yields as high as 80% from cultured human lymphoid cells. In contrast to low-frequency sound application, there is considerable destruction of the internal cytoarchitecture, with subsequent release of nuclear DNA and nucleoproteins. However, KCl seemingly liberates less lipoproteins from cultured human lymphoid cells than does low-frequency sound treatment. This is somewhat advantageous, as lipoproteins seem to interfere considerably with the purification of the active antigenic principle, possibly because of aggregation phenomena and occlusion of the antigenic moiety (Reisfeld et al., 1971b). The major advantage of the 3 M KCl method is its simplicity and reproducibility as well as its efficacy in producing relatively high antigen yields. The method has now found additional wide use for (1) the solubilization of HL-A antigens from spleen cells (Etheredge and Najarian, 1971) and from platelets (Uhlenbruck et al., 1973), (2) the extraction of soluble H-2 antigens from peripheral murine leukocytes (Götze and Reisfeld, 1972) and from cultured murine leukemia (L1210) cells (Götze and Reisfeld, 1972; J. H. Pincus, personal communication), and (3) the solubilization of tumor-specific antigen of diethylnitrosamine-induced guinea pig sarcomas (Meltzer et al., 1971) and of sarcoma and malignant melanoma specific antigens from human tumors (W. Winters, personal communication).

The mode of action of KCl in solubilizing HL-A antigens from the cell surface is generally assumed to be that of a "chaotropic agent". In this regard, ions long ago were recognized by Hofmeister (1888) to differ in their efficiency to salt out globulins. The ions most effective in this regard cause the folding, coiling, and association of proteins, whereas less effective precipitants trigger unfolding and dissociation. These latter anions have been called "chaotropic" by Hamaguchi and Geiduschek (1962), who proposed that they break hydrophobic bonds mainly by their disordering effects on the structure of water. The action of chaotropic ions and the role of water structure and hydrophobic bonding in the stabilization of macromolecules have been discussed in an extensive review by Dandliker and de Saussure (1971).

The effectiveness of anions such as SCN^-, ClO_4^-, and I^- to function as chaotropes in breaking hydrophobic bonds of macromolecules was demonstrated by their ability to resolve microsomal enzymes and complexes of the mitochondrial electron transport system (Hatefi and Hanstein, 1968) and by their ability to dissociate selectively antigen–antibody complexes (Dandliker et al., 1967). Although Cl^- is a relatively ineffective chaotropic agent for resolving substances associated with intracellular membrane systems (Hatefi and Hanstein, 1968), it proved highly effective for the solubilization of membrane-associated HL-A antigens from cultured lymphoid cells. In contrast, ClO_4^- was not effective, since it destroyed the antigenic activity of these substances (M. A. Pellegrino and R. A. Reisfeld, unpublished observations).

The KCl extraction technique is carried out essentially as described by Reisfeld *et al.* (1971*a*). Cultured human lymphoid cells are collected by centrifuging the cellular suspension at $400 \times g$ for 10 min at room temperature. If large amounts of cells are used, they are most efficiently collected by a continuous flow system (Sorvall Szent-Gyorgyi and Blum) with the flow rate adjusted at about 400 ml/min.

The cells are then washed three times with 0.15 M phosphate, 0.14 M saline at *p*H 7.4, in order to remove as much residual fetal calf serum as possible. The washed cells are dispersed by vibration on a Vortex mixer and resuspended in 0.9% sodium chloride containing 3 M KCl buffered with 0.015 M phosphate at *p*H 7.4 (20 ml solvent per 10^9 cells). This suspension is gently agitated on a mechanical shaker for 16 hr at $4\,^\circ$C and then centrifuged at $163,000 \times g$ (average) for 1 hr. The supernatant, when dialyzed at $4\,^\circ$C against two changes of 100 vol each of saline, forms a gelatinous precipitate which contains primarily DNA. This material is easily removed by centrifugation at $1500 \times g$ for 20 min. DNA could not be detected in the residual supernatant by the diphenylamine assay of a hot trichloroacetic acid extract (Burton, 1956). Less than 1% of the total labeled DNA was found in the $1500 \times g$ supernatant by a more sensitive technique: radioactive analysis of the dialyzed, centrifuged extract obtained from WI-L2 cells grown for two generations in the presence of thymidine-2-^{14}C (Reisfeld *et al.*, 1971*a*).

Experiments carried out to determine optimal extraction conditions

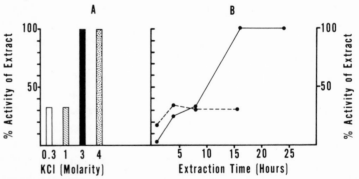

Fig. 1. KCl solubilization of HL-A antigens. A: Effect of molarity of KCl solution on antigen yield. B: Effect of time of extraction on antigen yield at $4\,^\circ$C (●————●) and at $40\,^\circ$C (●- - - - -●).

showed that 3 M KCl was more effective for solubilizing HL-A antigenic activity than either 0.3 M or 1.0 M KCl and that a 16-hr extraction time resulted in optimal yield (Reisfeld et al., 1971a) (Fig. 1). Prolonged extraction time, up to 96 hr, failed to increase the yield to any marked extent, but no destruction of solubilized HL-A antigenic activity was observed. Additional extractions of the 163,000 × g pellet either by 3 M KCl or by low-frequency sound solubilized limited amounts of proteins with good immunological activity and with the identical complex electrophoretic pattern as that of the 163,000 × g supernatant obtained from the original extract. An initial extraction of whole lymphoid cells for 1–2 hr followed by an additional 16-hr extraction of a 10,000 × g pellet of the first extract showed essentially similar results. However, when the KCl extraction procedure is carried out at 40°C, a considerably lower antigen yield is obtained than at 4°C (Reisfeld et al., 1971a).

Good viability of cultured cells utilized for antigen extraction is crucial in order to obtain maximal yields, as batches of cells with less than 50% viability yielded only about 10% soluble HL-A antigen with relatively poor immunological potency, in contrast with the 50–80% recovery values achieved when the cells showed 95–100% viability. As shown in Table I, extractions of the same cell line on different occasions resulted in antigenic preparations which varied with respect to total protein content, immunological potency as detected in the inhibition test, and consequently total amount of antigenic units recovered.

It should be pointed out that the antigen yield ratio among the phenotypic HL-A specificities of a given cultured cell remains essentially constant, independent of the overall antigen yield, when extractions are carried out on separate occasions. Furthermore, there is a good correlation between the ratio of the absorbing capacity of different HL-A specificities on the same cell and the blocking capacity of these same specificities present on HL-A antigens solubilized from these cells. This indicates that during the 3 M KCl extraction no preferential extraction or selective destruction of HL-A specificities takes place.

When various cell lines were analyzed, there was no direct relationship between relative density of HL-A determinants on lymphoid cell surfaces and serological activity of solubilized HL-A antigens (Pellegrino et al., 1973a). As also shown in Table I, upon analysis of two cultured lymphoid cell lines which showed an equal density of HL-A2 determinants, one line (RPMI 1788) yielded excellent recoveries of HL-A2 specificity, whereas much lower recoveries were obtained from the other line (RPMI 7249).

Table I. Soluble Antigen Yields from Various Human Cultured Cell Lines

Cell line	Antigen batch No.	HL-A1		HL-A2		HL-A3		HL-A5		HL-A7	
		ID_{50} u[a] 10^9 cells	Percent recovery[b]	ID_{50} u 10^9 cells	Percent recovery	ID_{50} u 10^9 cells	Percent recovery	ID_{50} u 10^9 cells	Percent recovery	ID_{50} u 10^9 cells	Percent recovery
RPMI 1788	1	—	—	620,000	51	—	—	—	—	165,000	30
	2	—	—	920,000	85	—	—	—	—	341,000	64
	3	—	—	462,000	38	—	—	—	—	170,000	31
	4	—	—	450,000	37	—	—	—	—	162,000	29
WI-L2	1	—	—	2,785,712	91	—	—	793,912	81	—	—
	2	—	—	1,800,000	60	—	—	521,371	53	—	—
RPMI 4098	1	—	—	—	—	1,666,666	100	—	—	—	—
	2	—	—	—	—	1,666,666	100	—	—	—	—
	3	—	—	—	—	833,333	50	—	—	—	—
RPMI 7249	1	350,000	17	350,000	28	—	—	—	—	35,000	12
	2	350,000	17	350,000	28	—	—	—	—	35,000	12

[a] Units.
[b] Percent recovery = $100 \times (ID_{50}$ units/10^9 cells)/(AD_{50} units/10^9 cells).

It is, of course, possible that the cultured cells of one of these lines may actually have more antigenic determinants on their surfaces than those revealed by quantitative absorption. Thus there could be masked sites not available for reaction with cytotoxic alloantibody. Nevertheless, such masked determinants may be readily extractable by 3 M KCl and would therefore contribute to higher antigen yields. Furthermore, the amount of soluble HL-A antigen extracted from cultured cells varies significantly with the stage of their growth cycle; antigenically more active preparations were obtained with cultures in late log or resting phase than with those in early log phase (Pellegrino et al., 1973a). As mentioned above, the density of HL-A determinants apparently does not contribute to the yield per se, since these determinants remain at essentially equal density throughout the cell growth cycle (Pellegrino et al., 1972b; Ferrone et al., 1973b).

Cell viability per se cannot account for the difference in antigen yield between cultures in early log phase and those in resting phase, since the percentage of dead cells, as determined by trypan blue uptake, is negligible in both cases. On the contrary, marked increase of dead cells in the cultures held in resting phase for several days could cause a reduction in the amount of HL-A antigen extracted, especially since dead cells have been shown to be a poor antigen source whenever 3 M KCl is utilized for their extraction. Serological data suggest that not all histocompatibility determinants on the cell surface are equally reactive, as some are apparently masked. Thus in vitro treatment of human normal lymphocytes with enzymes (Mittal et al., 1968; Yunis et al., 1970; Grothaus et al., 1971; Gibofsky and Terasaki, 1972; Braun et al., 1972) or with sulfhydryl compounds (Sirchia and Ferrone, 1971; Mercuriali et al., 1971) renders them more reactive in the cytotoxic reaction. The variation in the amount of "masking substance," e.g., mucopolysaccharides, and its spatial relationship on the cell surface to HL-A determinants might make the latter more or less available to the extraction procedure. In this regard, it has been shown that both the composition of the cell membrane and the turnover of its components vary during the cell growth cycle, being low in log phase and high in resting phase (Warren and Glick, 1968). An influence of the cell metabolism is also suggested by the fact that no variability in yield occurred when antigens were extracted from spent culture medium, i.e., where they are attached to small nonmetabolizing cell fragments (M. Pellegrino and R. A. Reisfeld, unpublished results) (see Table II). The existence of a pool of histocompatibility antigens inside the cells has not yet been established. Thus Haughton (1966) could not show any increase in the absorbing capacity of H-2 alloantisera which were reacted with mouse

Table II. Solubilization of HL-A2 Specificity from Lymphoid Cells and from the
Exhausted Medium of Cultures in Different Stages of Cell Growth

Source	ID_{50} (HL-A2) units/10^9 cells			
	0.8^a	1.90^a	2.46^a	1.80^a
Cells	280,000	2.050,000	1,555,000	19,922
Medium	250,000	233,324	312,500	300,000

[a] Cells/ml $\times$ 10^6.

cells disrupted by sonication. On the other hand, Manson and Palm (1972)
reported that disrupted L5178Y cells had higher immunogenic potency for
H-2 antigens, suggesting the presence of histocompatibility antigens inside
the cells. It is indeed difficult to resolve the question of the existence of an
intracellular antigen pool by such experiments. First, one cannot rule out
possible destruction of HL-A antigens during the cell disruption process,
and, second, the enhanced immunogenicity does not necessarily reflect an
increased amount of histocompatibility antigens. If a pool of HL-A antigens
does indeed exist within the lymphoid cell, a variation in its size during the
cell growth cycle might account for the observed differences in amounts of
solubilized HL-A antigens.

It is interesting to note that only small yields (3–8 %) of soluble HL-A
antigen could be obtained by KCl treatment of crude cell membrane preparations obtained by freezing and thawing as well as from lyophilized cells
(Reisfeld et al., 1971a; M. A. Pellegrino and R. A. Reisfeld, unpublished
observations). In addition, when KCl was utilized with membranes prepared
according to Mann et al. (1969) or alternatively by utilization of the detergent
Nonidet P-40 (NP-40) as well as by a method applying fixation by zinc ions
according to Warren et al. (1966), uniformly poor yields (1–3 %) of HL-A
antigen were obtained (M. A. Pellegrino and R. A. Reisfeld, unpublished
observations). However, when materials judged to be membrane-like by
electron microscopy were recovered from spent culture medium and extracted
by the 3 M KCl method, higher recoveries than those mentioned above were
obtained in the soluble extract, in some cases reaching 76 % of the serological
activity displayed by the cell membrane–like materials (M. Pellegrino and
R. A. Reisfeld, unpublished results).

Crude extracts obtained either from whole cultured cells or from spent
culture medium have proven to be very stable at −20 °C for more than 1

year. However, in a few instances, antigen prepared on different occasions from the same cell line did show various degrees of instability upon prolonged storage. Out of ten preparations with the same HL-A2 activity studied over a 1-year period, five completely retained their initial serological activity while four others lost about 50%, with one sample being completely devoid of any remaining activity. The decrease of serological activity follows two patterns: one a slow but continuous decline and the other a sudden decrease after a certain period of stability. All tests in these experiments were performed with small aliquots frozen and thawed only once. However, a limited number of freezings and thawings of small aliquots (up to ten times) does not seem to affect the activity of these KCl extracts, although some material is precipitated on each occasion.

Furthermore, filtering the crude 3 M KCl extract through Millipore filters (0.22 μ) at low N_2 pressure ($\sim$ 10 psi) renders the occasionally opalescent antigen crystal clear without altering any of its above-described properties as far as they can be detected by the serological inhibition assay.

2.5. Mechanism of Action of Chaotropic Ions

The effective solubilization of any substance from a cell membrane network poses a difficult problem because many membrane structures are held together primarily by strong hydrophobic bonds. Thus cell surface–associated substances often possess poor water solubility, and their lack of stability in aqueous media impairs the successful resolution of their molecular organization. In this regard, hydrophobic bonds are the most significant forces contributing to the integrity of cell membrane structures, mainly because van der Waals attractions between apolar groups are weak and hydrogen bonds of the $C = O \ldots H\text{-}N$ and $C = O \ldots H\text{-}O$ type remain thermodynamically unstable if not protected from water (Klotz and Farnham, 1968). It is largely because of this unfavorable thermodynamic reaction with water that apolar groups form hydrophobic bonds. A decrease in entropy accompanies the transfer of an apolar molecule from a lipophilic surrounding to water, an event thought to be related to the highly ordered structure of water. However, increasing any disorder in water structure can diminish this negative entropy change. Such a condition can be effected by some inorganic anions, e.g., SCN^-, ClO_4^-, I^-, Br^-, and Cl^-, as they possess large positive entropies due to their structure-breaking effects on water. Such anions also decrease the polarity of the surrounding water, rendering it more lipophilic. Thus hydrophobic bonds responsible for native membrane structures are weakened, and entry of apolar groups in the aqueous phase is facilitated (Kauzman, 1959). Consequently, chaotropic ions can be highly effective in rendering cell membrane–associated substances, viz., HL-A antigens, soluble in aqueous solvents.

The mode of action of KCl in the extraction of water-soluble HL-A antigens is generally assumed to be that of a chaotropic agent dissociating noncovalent hydrophobic associations, as discussed above. It seems relatively unlikely that enzymic action affecting covalent bonds would play a significant role during this solubilization procedure, i.e., in the presence of 3 M KCl. In the unlikely event of a significant participation by intracellular enzymes in the solubilization of HL-A antigens, the most likely candidates would be lysosomal enzymes, viz., cathepsins. In this regard, native proteolytic enzymes of white cells, cathepsins A, C, D, and E, have pH optima ranging from 2.5 to 5.0 (Greenbaum, 1971). The pH maintained during KCl extraction of cultured lymphoid cells, pH 7.4, seems not to be conducive for maximal reactivity of these enzymes. For these enzymes to be truly instrumental in antigen solubilization, one would expect them to be equally effective at 1.0 and 3 M KCl, since under these conditions there is an equal destruction of intracellular cytoarchitecture. However, this is not the case, since there is considerably less antigen yield at 1.0 KCl concentration. The molarity of KCl is obviously very important to effect optimal dissociation of hydrophobic associations. Another argument against effective solubilizing effects of catheptic enzymes at 4 °C is, as mentioned above, the fact that even prolonged extraction (up to 96 hr) does not decrease the specific activity of solubilized HL-A antigens. In this regard, prolonged digestion (in excess of a few hours) of lymphocytes with proteolytic enzymes, e.g., papain, causes complete loss of any residual antigenic activity (Shimada and Nathenson, 1969). Although the inadvertant participation of intracellular proteases can never be completely ruled out in any salt extraction of cultured human lymphoid cells, for the reasons cited above we feel that such enzymes may play only a very minor role, if any at all, in the 3 M KCl extraction of soluble HL-A antigens from cultured human lymphoid cells.

3. PURIFICATION OF HL-A ANTIGENS

To isolate and purify a biologically active substance from cell membranes in water-soluble form poses a formidable problem for a number of reasons. As mentioned above, it is difficult to render these antigens soluble due to their intimate association with the cell membrane; furthermore, the non-selectivity of the extraction procedure results in vast amounts of contaminant substances associated with very small amounts ($\sim$ 1%) of solubilized HL-A antigen. The task of purification of the antigenic moiety is further complicated by the complexity of the *in vitro* assay system (see below).

As far as the purification strategy is concerned, it would seem logical to apply as many different fractionation principles as feasible, including fractional salting out, gel exclusion chromatography, ion-exchange chromatography, and electrophoresis. Such an approach seems especially warranted considering the large body of experience with naturally soluble enzymes or serum proteins. However, our experience with these cell membrane-associated antigens dictated another approach to the purification problem, especially since the biological activity of HL-A antigens is lost by extensive manipulation. This has indeed been confirmed by other workers (Mann *et al.*, 1968, 1969) and also by the finding that following an extensive, multistage purification protocol as much as 98% of the biological activity present in crude, papain-solubilized murine H-2 antigen extracts was lost and the product still did not show electrophoretic homogeneity (Shimada and Nathenson, 1969). Consequently, most concepts derived from purification of soluble serum proteins do not apply to histocompatibility antigens, which are difficult to solubilize because of their membranous association with hydrophobic substances. Purification of these materials is further complicated by the fact that their "solubility" may be in a precarious balance, and a number of factors may render these antigens less soluble at any time during the purification process. For example, changes in solubility may occur because of sudden removal of other components of the isolate or because of alterations in the local environment possibly resulting from absorption-desorption or alteration in ionic strength or in charge environment during chromatography or discontinuous gradient electrophoresis. In view of these problems, the best strategy seems to be the application of a single procedure affording maximal resolution. This criterion has been met, at least in part, by preparative acrylamide gel electrophoresis (PAGE) in a discontinuous buffer system, a method which at present is the procedure of choice for the purification of H-antigens.

3.1. Preparative PAGE

PAGE has found wide application since its introduction (Raymond and Weintraub, 1959; Ornstein, 1964), largely because of its high resolving power, sensitivity, and versatility. The method has been used mainly as an analytical tool, with qualitative patterns being used for the interpretation of results. Recent developments, notably those related to the mathematical parameters of the physical–chemical nature of PAGE, resulted in a computer program which generated 4000 discontinuous buffer systems between pH 3 and 11 (Rodbard and Chrambach, 1971). This advance together with the standardization of polymerization and electrophoresis conditions has greatly improved the versatility and reproducibility of PAGE (Rodbard and Chrambach, 1971; Rodbard et al., 1971; Lunney et al., 1971).

Zone electrophoresis in polyacrylamide gel is quite effective when one efficiently utilizes this array of buffer systems while simultaneously exploiting differences in size and charge of the protein moieties to be resolved. PAGE is largely based on the achievement of a high degree of pore size reproducibility (over the range 0.5–3 nm) by an adjustment of the total concentration of acrylamide and its cross-linking agent under carefully standardized conditions (Rodbard and Chrambach, 1971; Rodbard et al., 1971). The charge separation between two macromolecules can be maximized when the net charge(s) of the molecules of interest and the contaminant(s) are known. This approach is exploited when expressing the electrophoretic data in "Ferguson plots"; i.e., the log of mobility vs. gel concentration. This plot yields via computer methodology information on molecular size, free mobility, and valence of protein molecules (Rodbard et al., 1971; Lunney et al., 1971).

Preparative PAGE has been successfully applied to isolate HL-A antigens from crude extracts obtained by 3 M KCl extraction of cultured human lymphoid cells (Reisfeld and Kahan, 1970b; Reisfeld et al., 1971a). It is feasible to apply as much as 100 mg of crude antigenic extract to a Buchler Polyprep 100 column and obtain approximately 1–2 mg of highly purified HL-A antigen. In order to purify HL-A antigens by preparative PAGE, soluble extracts in saline are dialyzed against a buffer at pH 6.7 (0.045 M tris, 0.032 M H_3PO_4), and 100 mg of protein is applied to the Polyprep column. Electrophoresis is performed at 0 °C in system "B" of Rodbard and Chrambach (1971) (pH 9.6, $7^1/_2$ % acrylamide gel) at a constant current of 35 mA. Electrophoretic resolution of HL-A antigenic extracts is optimal at pH 9.6, as determined by analytical PAGE at various pH values (pH 3.56, 6.96, 7.35, 7.67, and 9.6). Fractions (8 ml) are collected at a flow rate of 0.8 ml/min with a tris–HCl elution buffer (0.138 M Tris, 0.18 M HCl, pH 8.2, containing 5 % sucrose). The fractions eluting at R_f 0.78–0.80 exhibit specific antigenic activity. Figure 2 is a typical elution diagram. Upon re-electrophoresis in the same preparative PAGE system, the active

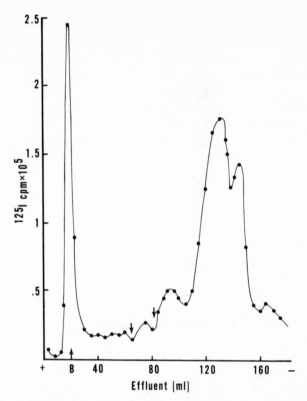

Fig. 2. Effluent diagram following preparative acrylamide gel electrophoresis of KCl extract from cell line RPMI 1788. The arrows indicate the region of antigenic activity.

antigenic moiety consists essentially of a single electrophoretic component (Fig. 3).

To assess protein yield and recovery of antigenic activity from preparative PAGE, a radiolabel method was combined with Kjeldahl nitrogen

Table III. Purification of HL-A Antigens by Preparative Acrylamide Gel Electrophoresis

Antigen	mg protein / 10^4 cells	ID_{50} units / mg protein	Total ID_{50} units
KCl extract	25	6,640	166,000
R_f 0.78 fraction	0.56	133,333	74,666

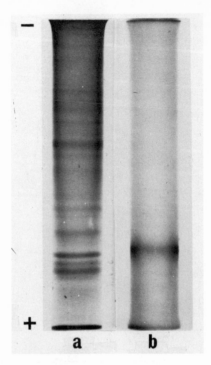

Fig. 3. **Electrophoretic patterns.** (a) KCl extract; (b) Antigen preparation obtained following reelectrophoresis of a preparative acrylamide fraction with R_f 0.78–0.80.

analysis. For this purpose, 1×10^9 cultured cells in log growth phase (2×10^7 cells/ml) were incubated for 4 hr with a mixture of ^{3}H-labeled amino acids (2.5 mCi total) in Eagle's minimum essential medium containing only 1% of its normal amino acid content. Under these conditions, approximately 15% of the radiolabel was incorporated into the cells, while 2% was found in the ultracentrifugal supernate and 0.04% in the purified antigen. Table III shows yield and activity of the purified HL-A antigen. In the best case, 2% of the nitrogen and from 45 to 60% of the total antigenic activity applied to the column initially can be recovered in the purified component with a twentyfold increase in ID_{50} units/mg nitrogen.

3.1.1. Advantages and Disadvantages

It should be pointed out that although the above-described purification procedure has worked for us, it may not represent the optimal approach to purify HL-A antigens. In this regard, isotachophoresis utilizing synthetic polyamino-polycarboxylic aliphatic ampholytes as "spacers" between protein components in a stack is being considered. In particular, it seems that a preparative application of this technique may be promising, as it utilizes continuous-flow isotachophoresis on a bed of Sephadex beads (W. Fawcett, personal communication). Employing a supportive medium other than polyacrylamide is most worthwhile, as polyacrylamide poses a number of problems particularly as far as irreversible adsorption of some proteins is concerned. Thus, while some plant proteins such as a mitogenic substance isolated from *Phytolacca americana* were purified most effectively by preparative PAGE (Reisfeld *et al.*, 1967), others, e.g., rabbit Ig light polypeptide chains (Reisfeld, 1967), were considerably more difficult to resolve. As far as HL-A antigens are concerned, the preparative PAGE method is thus far the only effective way to purify these substances. This technique is, however, far from optimal for this purpose, as it recovers at best 60% of the original activity units from a given extract. However, it should be pointed out that we demand a great deal from this method in the purification of HL-A antigens. For example, the extract applied to the polyacrylamide gel column is chemically very complex, containing simple protein moieties as well as more complex lipo- and glycoproteins and a variety of DNA and RNA moieties. As pointed out above, partial purification by fractional salting out, gel exclusion, and ion-exchange chromatography is indeed possible. However, the price which has to be paid for this is high indeed, as often only very limited purification and excessively high losses of biological activity of the antigen result from all these additional manipulations. The obvious answer to this problem plagueing the H-antigen field is the development of a more selective solubilization method which avoids the vast amounts of contaminant materials in the initial antigenic extract. In order to design such a method, it is, of course, necessary to know as much as possible not only about the biological properties of HL-A antigens but also about their chemical characteristics. In this regard, our overall approach carried out together with a thorough biological and serological characterization of HL-A antigens was designed with a dual purpose in mind: (1) to evaluate thoroughly the chemical properties and to gain an understanding ol the molecular basis of HL-A cell surface antigens and their immunologicaf function and (2) to be able to apply this chemical information to design more selective and specific procedures to isolate these biological cell surface markers in soluble form.

4. PHYSICOCHEMICAL CHARACTERIZATION OF HL-A ANTIGENS

4.1. Physical Properties

During the past decade, there have been considerable efforts to elucidate the chemical nature of histocompatibility antigens. However, progress has been relatively slow due in part to the chemical complexity of these cell surface located substances and the lack of adequate methods to purify them. Thus substances as diverse as DNA, lipid, carbohydrate, and protein in various conjugated forms have been, at one time or another, implicated as the antigenic principle (for review, see Kahan and Reisfeld, 1969a; Reisfeld and Kahan, 1970a).

HL-A antigens extracted by either low-frequency sound or 3 M KCl possess size homogeneity. The molecular weight of antigen solubilized by low-frequency sonication was 34,600 daltons by Yphantis sedimentation equilibrium analyses, assuming a partial specific volume of 0.72. The material was 94% monodisperse, containing 6% of an aggregated moiety with a molecular weight of 150,000 (Kahan and Reisfeld, 1969b). The antigen solubilized with 3 M KCl is essentially monodisperse and has a sedimentation coefficient $S_{20,w}$ of 2.3 and a molecular weight of 31,000 daltons by Archibald sedimentation equilibrium analyses, assuming a partial specific volume of 0.72. Their similar molecular weights and practically identical electrophoretic mobilities suggest that a single molecular antigenic entity on the cell membrane is derived by these two methods which rely mainly on noncovalent bond dissociation.

The electrophoretically purified antigen preparations not only possess size uniformity, but they also show electrophoretic homogeneity in polyacrylamide gels at varying pH values and gel porosity. Structural homogeneity was also suggested by the amino acid compositions approaching whole integer residue values and by tryptic peptide maps showing the number of peptides (24) expected from the arginine (5) and lysine (18) residues found upon amino acid analysis (Reisfeld and Kahan, 1972b).

4.2. Chemical Properties

Although, as mentioned above, practically every chemical entity known to man was at one time or another held to be the antigenic principle of H-antigens, there is now a general consensus in the field that this principle is polypeptide in nature. Available chemical data strongly support this hypothesis. Thus antigenic activity is irreversibly destroyed by proteolytic enzymes, protein denaturants, extreme pH values (above 10 and below 4), temperatures greater than 50 °C, or detergents and complex salts in concentrations sufficient to affect protein conformation (for reviews, see Kahan and Reisfeld, 1969a; Reisfeld and Kahan, 1970a). Although repeatedly postulated, there has been relatively little evidence suggesting that carbohydrate moieties are essential for antigenic activity, even though papain-solubilized H-2 and HL-A antigens were reported to contain from 3 to 8 % neutral carbohydrates (Sanderson et al., 1971). However, digestion of these antigens with carbohydrases, which split off much of the neutral carbohydrates and hexosamines, or with neuraminidase, which cleaves off neuraminic acid residues, has essentially no effect on the antigenic activity of H-2 antigens (Muramatsu and Nathenson, 1971). In this regard, it is also of considerable importance that both highly purified guinea pig transplantation antigens and HL-A antigens solubilized by methods effecting essentially noncovalent bond cleavage do not contain either lipid or carbohydrate at levels greater than 1 %. Thus guinea pig antigen (molecular weight 15,000) and HL-A antigens (molecular weight 31,000) contain, respectively, at best one or two carbohydrate residues per mole.

One of the more striking findings which strongly supports the concept of antigenic determinants being protein in nature is the reproducible and significant differences found among the amino acid compositions of antigens extracted from cultured cell lines derived from histoincompatible individuals. Differences in the content of serine, alanine, valine, leucine, and isoleucine were detected between antigens derived from strains 2 and 13 of histoincompatible, inbred guinea pigs (Kahan and Reisfeld, 1968). In addition, amino acid analyses of highly purified HL-A antigens solubilized by low-frequency sound from cell lines possessing differing alloantigenic specificities showed differences in the content of aspartic acid, serine, proline, alanine, and tyrosine (Reisfeld and Kahan, 1972a). The remaining amino acids were present in strikingly similar amounts. It seems reasonable to suggest that, as demonstrated with immunoglobulin light chains with different allotypic specificities, these HL-A antigens also possess genetically segregating alloantigenic specificities which are expressed by polypeptide structure.

As initially reported, the amino acid composition of soluble HL-A alloantigens showed little or no half-cystine or methionine (Kahan and Reisfeld, 1969b). However, these analyses were not carried out following either extensive reduction and carboxamidomethylation in guanidine or performic acid oxidation of the antigen preparations. More recent work, where such procedures were employed, revealed four residues of cysteic acid and 3.6 residues of S-carboxymethylcysteine, indicating the presence of four half-cystine groups (Reisfeld and Kahan, 1972b).

5. MOLECULAR REPRESENTATION OF HL-A ANTIGENS ON THE LYMPHOID CELL SURFACE

The manner by which histocompatibility antigens are attached to lymphoid cell membranes may be elucidated by an analysis of the products obtained by covalent and noncovalent methods as well as by the kinetics of their release. In this regard, the discrepancy between the carbohydrate content of HL-A antigens released by proteolytic digestion and by salt extraction methods is most likely related to the membrane sites attacked by these two techniques. The noncovalent method yields an antigenic moiety with a molecular weight of 31,000 containing essentially no carbohydrate, whereas proteolysis yields fragments of 40,000–75,000 as estimated by Sephadex gel exclusion chromatography containing from 3 to 8% carbohydrate. It seems indeed possible that the glycoprotein fragments released by digestion with a nonspecific proteolytic enzyme such as papain represent a membrane unit consisting, at least in part, of the antigenic determinant combined with glycolipid and other surface proteins, thereby yielding the larger-size fragment.

There also is a difference in the kinetics of antigen release as effected by covalent and noncovalent procedures. Papain solubilizes H-2 and HL-A antigens within minutes and then irreversibly denatures them if allowed to interact in excess of 60 min at 37°C (Shimada and Nathenson, 1969). In contrast, although sonication is capable of solubilizing HL-A antigens within minutes, it is unable to denature the antigenic moiety even after prolonged exposure (Reisfeld and Kahan, 1970b). The solubilizing action of 3 M KCl is relatively slow, as it requires hours; however, the treatment can be continued for up to 4 days at 4°C without detectable loss of specific antigenic activity. Since determinants of histocompatibility antigens are highly susceptible to proteolysis while apparently not being affected by dissociation of noncovalent bonds, it seems likely that the antigenic determinants depend mainly on primary structure rather than on extensive three-dimensional conformation.

Another distinct difference between the covalent and noncovalent methods of antigen release may reflect the mode of attachment of these substances on cell surface membranes. Papain apparently can act on susceptible peptide bonds and readily releases HL-A antigens even from large amounts of structurally denatured (frozen-thawed) cell membranes. In contrast, noncovalent techniques, i.e., low-frequency sound and 3 M KCl extraction, can liberate antigens effectively only from fresh nondenatured cell membranes. In addition, membranes from frozen-thawed cells contain only 35% of the absorptive capacity of fresh, intact cells. This would suggest that freeze-thawing alters the membrane structure in such a way as to obscure regions containing antigenic determinants, trapping them within denatured lipoprotein lattices which are then susceptible only to proteolytic attack.

It seems likely that the membrane unit bearing histocompatibility determinants is associated with the polar ends of lipoprotein lattices by hydrophobic bonds with apolar groups. Since noncovalent techniques readily release antigens, the molecular antigenic entities are probably noncovalently attached to a complex lipid–lipoprotein matrix. Although highly speculative, it is tempting to propose a dynamic view of HL-A antigens with an active turnover of some areas of the cell membrane corresponding to the previously mentioned "patches" visible upon ultrastructure analysis. Thus while some "patches" are being shed, new ones are being assembled. However, more evidence on biosynthesis and turnover of HL-A antigens is required before such a hypothesis can be justified.

5.1. Membrane–Antigen Models

Cellular membrane structure according to the traditional model of Danielli and Davson (1935) is postulated to consist of phosphatide bilayers, the surfaces of which are coated with membrane proteins linked to lipid head groups by ionic and/or hydrogen bonds. Hoelzl-Wallach and Gordon (1969) do not believe this traditional view to be compatible with more recent data suggesting that membranes represent complex mosaics of organized assemblies of lipoproteins with certain architectural organizations. Accordingly, membrane proteins are believed to be located on membrane surfaces and also within the apolar core of the membrane. These investigators propose that the intact cell surface is not uniform but that it is made up of discrete patches with surface areas of 1.2×10^6 Å^2 and that these patches fall into a number of groups with distinct protein compositions. More recently, Nicolson et al. (1971), using an indirect ferritin-labeling technique, found that H-2 alloantigenic sites on mouse erythrocyte membranes showed an irregular, patchlike distribution. On the basis of these observations and those of a number of other investigators studying cell surface antigens, as well as thermodynamic considerations, Singer and Nicolson (1972) extended the somewhat more static membrane models mentioned above, suggesting that a fluid mosaic model is consistent with the restrictions imposed by thermodynamics. Globular molecules are thought to be partially embedded in a phospholipid matrix, which in turn is considered to be a discontinuous fluid bilayer. Accordingly, this fluid mosaic is thought to be "analogous to a two-dimensional oriented solution of integral proteins (or lipoproteins) in the viscous phospholipid bilayer solvent." The experiments of Frye and Edidin (1970) led Singer and Nicolson (1972) to postulate that the fluid mosaic model of membrane structure is "the result of the free diffusion and intermixing of the lipids and the proteins (or lipoproteins) within the fluid lipid matrix." In their experiments, Frye and Edidin (1970) used Sendai virus to produce heterokaryons; i.e., they fused cultured human and murine cells. An indirect fluorescent technique using various labels, and H-2 and HL-A alloantisera showed that after 40 min at 37°C the histocompatibility determinants were for the most part completely intermixed. This result suggested to Frye and Edidin (1970) that intermixing of membrane components can be attributed to their diffusion within the membrane rather than to their removal and reinsertion or possible synthesis and reinsertion into the heterokaryon membrane. As pointed out by Singer and Nicolson (1972), these experiments apply only to those components on the cell surface readily available to antibody reagents and do not answer the question as to whether all components on the surface, especially those on inner cell membrane surfaces, do indeed intermix and diffuse. At any rate, until there exists sufficient experimental evidence to disprove it, the fluid mosaic model of cell membranes seems to be quite useful to establish membrane–antigen models which hopefully can be tested experimentally.

6. SEROLOGICAL ASSAYS

6.1. The Cytotoxic Test

Among many methods developed for the study of histocompatibility antigens on the surface of lymphoid cells, the complement-dependent cytotoxic test is the most widely used, especially after the miniaturization of the assay by Terasaki and McClelland (1964). The test relies on an alteration of the permeability of the cell membrane induced by action of complement on cells coated by cytotoxic antibodies. The exclusion of supravital dyes, such as eosin (Terasaki et al., 1967) and trypan blue (Walford et al., 1964; Engelfriet and Britten, 1965; Amos, 1966), is the most frequently used procedure to detect cells killed by cytotoxic antibody and complement.

The assay system found most effective in this laboratory is a modification (Pellegrino et al., 1972c) of the microdroplet technique described by Mittal et al. (1968). This reaction is performed at room temperature under mineral oil in disposable chambers (Möller-Coates, Moss, Norway), which permit a homogeneous dispersion of cells in the microdroplets and thus facilitate a more quantitative reading. Human cytotoxic alloantisera (2 μl) are mixed with target leukocytes suspended in Hank's balanced salt solution (BSS) (1 μl containing 2000 cells); after a 30 min incubation, selected rabbit complement (3 μl) is added, and the mixture is incubated for an additional 60 min. Then eosin (5%, 2 μl) is added to each microdroplet, and after 2 min the reaction is stopped by the addition of 1 μl of 36% formalin. The microdroplets are examined for cell survival by inverted phase contrast microscopy at a magnification of 320 $\times$.

The target cells are human peripheral lymphocytes or cultured lymphoid cells; the latter are more sensitive in the cytotoxic test than the former, reflecting at least in part the higher density of HL-A determinants on the cell surface (Table IV) (Rogentine and Gerber, 1970; Pellegrino et al., 1972a). No data are available concerning the sensitivity of these two types of cells to the lytic action of complement; however, it seems worthwhile to mention that cultured lymphoid cells activate the early components of complement sequence in absence of antibody, while peripheral lymphocytes do not (Ferrone et al., 1973b).

Table IV. AD_{50}[a] Values of Peripheral Leukocytes and Human Cultured Lymphoid Cells Derived from the Same Donor

Donor	Cell type	AD_{50}			
		HL-A2	HL-A3	HL-A5	HL-A7
B. P.	PL[b]	2,800	> 50,000	> 50,000	3,800
	CL[c]	825	> 50,000	> 50,000	1,800
D. O.	PL	3,200	> 50,000	> 50,000	23,000
	CL	800	> 50,000	> 50,000	3,500

[a] Number of cells required to reduce by 50% the cytotoxicity of HL-A alloantisera.
[b] Peripheral leukocytes.
[c] Cultured lymphoid cells.

HL-A alloantisera are obtained from either multiparous women, polytransfused patients, recipients of organ transplants, or deliberately immunized subjects. As stressed by Walford *et al.* (1967), these alloantisera are only operationally monospecific, as most of them contain additional, although relatively weak, cytotoxic antibodies which may become lytic whenever the sensitivity of the test system is increased. This possibility should always be kept in mind in HL-A typing of cultured lymphoid cells, since typing alloantisera have been characterized strictly on the basis of their reactivity with peripheral lymphocytes, which are much less sensitive than cultured lymphoid cells in the cytotoxic test. It is strongly recommended that the results of typing cultured lymphoid cells be verified by the direct cytotoxicity test, especially when discrepancies are detected between the HL-A phenotype of cultured cells and of the peripheral lymphocytes from which the cell line was derived (Rogentine and Gerber, 1970). This procedure will also detect any CYNAP (cytotoxicity negative–absorption positive) reaction (Ferrone *et al.*, 1967), which can make the interpretation of the serological evaluation of soluble HL-A antigens considerably more difficult.

Complement is a pool of rabbit sera. Among several sources of complement tested, rabbit serum has proven to be the most efficient in the cytotoxic reaction (Walford *et al.*, 1964), although it has a relatively low level of complement components when measured in the conventional hemolytic test system with sensitized sheep erythrocytes (Nelson and Biro, 1968). The superiority of rabbit complement in the lymphocytotoxic test is based on the fact that rabbit serum contributes to the reaction complement components and natural antibodies which enhance the complement-binding activity of HL-A alloantibodies (Ferrone *et al.*, 1971*a*). The titer and the specificity of the natural antibodies vary from rabbit to rabbit (Mittal *et al.*, 1973*b*); since these natural antibodies play a crucial role in the HL-A lymphocytotoxic test, it is advisable to use as source of complement a pool of sera from a large number of rabbits in order to minimize any source of variability arising from the different amounts and specificities of natural antibodies as well as from the level of complement components. When peripheral lymphocytes are the target cells, rabbit complement has to be selected for lack of spontaneous cytotoxicity to peripheral lymphocytes from four or five subjects of different ABO groups. When cultured lymphoid cells are the targets, rabbit sera must be either absorbed (McDonald *et al.*, 1970) or diluted (Rogentine and Gerber, 1969; Ferrone *et al.*, 1971*a*; Takasugi, 1971) to eliminate their direct cytotoxic effects on these cells. A rapid and convenient procedure followed in our laboratory is to dilute rabbit serum with

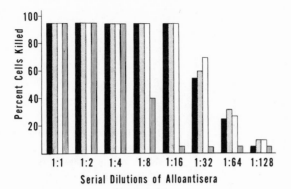

Fig. 4. Effect of various sources of complement (C) on titer of cytotoxic alloantisera (TO-11-03) with cultured human lymphoid cells WI-L2. Solid bar, unabsorbed rabbit C diluted 1:4 with human C; stippled bar, rabbit C absorbed with cultured cells; open bar, rabbit C absorbed with peripheral blood cells and diluted 1:4 with BSS; striped bar, human C.

AB human serum (Ferrone *et al.*, 1971*a*); the natural rabbit antibodies are blocked by human IgM (Herberman, 1970) and the mixture is as effective as rabbit complement absorbed with cultured human lymphoid cells (Fig. 4).

6.2. The Blocking Assay

The blocking assay is based on the ability of soluble HL-A antigens to combine specifically with HL-A alloantibodies and to prevent subsequent antibody activity against target cells in the complement-dependent lymphocytotoxic test.

Selected dilutions of alloantiserum are preincubated for 60 min at room temperature with 1 μl of twofold progressive dilutions of soluble HL-A antigens (Kahan *et al.*, 1968; Pellegrino *et al.*, 1972*c*). Then the target cells are added to the mixture, and the reaction is performed as described for the direct cytotoxic test. Decreased potency of alloantibody caused by its specific reaction with soluble HL-A antigens is detected as an increased percentage of viable cells. For each inhibition test, several control reactions must be performed; alloantiserum is preincubated with BSS instead of soluble antigen in order to check the serum titer. The viability of the target cells is examined by incubating them with BSS alone and complement alone; the experiment is discarded whenever the viability of the target cells is less than 95%.

The immunological potency of a soluble antigen preparation under

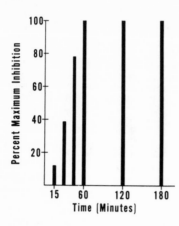

Fig. 5. Influence of incubation time of the alloantiserum with soluble HL-A antigen on the inhibition of the cytotoxic reaction. Alloantiserum
and soluble HL-A antigen were preincubated for
various periods of time before the addition of
target cells and complement.

investigation is expressed by its ability to inhibit the lymphocytotoxic
reaction. The percentage inhibition is calculated using the following formula:

$$\% \text{ inhibition} = 100 - \frac{\% \text{ cells killed in presence of soluble HL-A antigen}}{\% \text{ cells killed in absence of soluble HL-A antigen}} \times 100$$

The percentage by which the cytotoxic potency of a given HL-A alloantiserum
is inhibited can be plotted on an arithmetic scale against the amount (μg
protein/μl) of antigen necessary to be added to the antiserum to achieve this.
The relationship between these two parameters is expressed in sigmoidal
fashion; from this curve, it is possible to calculate the inhibition dosage
(ID_{50}), the amount of antigen required for a 50 % reduction of the cytotoxic
activity of the alloantiserum (Fig. 5). This parameter is currently used to
express the immunological potency of a soluble HL-A antigen preparation,
as will be discussed below.

Sixty minutes is the optimal incubation time of soluble HL-A alloantigens with anti-HL-A alloantisera in order to attain maximal inhibition
of cytotoxic antibodies; shorter times of incubation do not accomplish a
complete blocking of the cytotoxic activity of the HL-A alloantibody, while
increasing the incubation time does not improve the sensitivity of the test
(Fig. 6). Temperature ranges between 20 and 37 °C do not affect the test
results, but for practical reasons it is convenient to perform the blocking test
at room temperature.

Inhibition of the cytotoxic activity of operationally monospecific
HL-A alloantisera directed against specificities present on the lymphoid cells
used for antigen extraction is more effective when soluble HL-A alloantigens

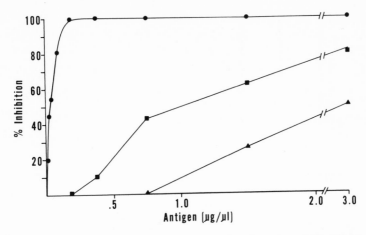

Fig. 6. Inhibition of cytotoxic activity of various alloantisera by soluble HL-A antigen extracted from cultured human lymphoid cells RPMI 1788 (HL-A2,10,7, W 14). ●, TO-11-03 (anti-HL-A2); ■, Grubisch 2-65-0-02-25-01 (anti-HL-A9); ▲, D66-6222-IV (anti-HL-A5). Alloantisera were employed at zero cytotoxic units.

are incubated with alloantisera rather than with target cells (Pellegrino et al., 1973b). However, the nonspecific inhibitory effect of several compounds on the lymphocytotoxic test is greater when these compounds are incubated together with target cells rather than with alloantisera (Hirata and Terasaki, 1972). The dissimilar results obtained in the two experimental systems suggest a different mechanism of inhibition; nonspecific inhibitors can bind to lymphocytes and thus inhibit their interaction with anti-HL-A alloantisera, whereas soluble HL-A alloantigens react specifically with anti-HL-A antibodies. The absence of an interaction between soluble HL-A antigens and target cells is also suggested by the observation that lymphocytes lacking an HL-A specificity, e.g., HL-A3, when incubated with large amounts of soluble HL-A3 antigen, do not acquire any sensitivity to anti-HL-A3 antibodies in the lymphocytotoxic reaction (Pellegrino et al., 1973b).

The solvents used to dilute soluble HL-A alloantigens can greatly influence the sensitivity of the blocking test (Pellegrino et al., 1973b); while BSS, fetal calf serum, or human albumin does not exert any detectable effect, glycine, due to its anticomplementary activity, effects a nonspecific inhibition of the cytotoxic reaction, and human serum causes an apparent increase in the serological activity of soluble HL-A alloantigen (Fig. 7). This last effect is caused by the inhibitory activity of human sera on rabbit complement (Ferrone et al., 1967), probably due to the interaction of human

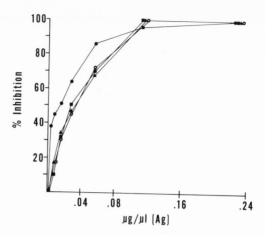

Fig. 7. **Effect of various diluents on the sensitivity of the inhibition assay.** Soluble HL-A alloantigen extracted from cultured human lymphoid cells WI-L2 was tested with antiserum TO-11-03 (anti-HL-A2). ●, AB serum, 10% v/v; ■, BSS; ▲, fetal calf serum, 10% v/v; ○, human serum albumin, 1% w/v.

IgM with natural rabbit antibodies (Herberman, 1970). These data indicate that any diluent used in the inhibition assay should be carefully tested to assess its effects. It should be emphasized that the agents used for extraction of soluble HL-A antigens such as KCl (Reisfeld *et al.*, 1971a), proteolytic enzymes (Mann and Fahey, 1971), cysteine (Chapel and Welsh, 1972), and 2-aminoethylisothiouronium bromide (M. A. Pellegrino, S. Ferrone, and R. A. Reisfeld, unpublished observations) are all capable of interfering with the lymphocytotoxic test. In fact, 3 M KCl can inhibit the formation of antigen–antibody complexes, while proteolytic enzymes (Gibofsky and Terasaki, 1972) and sulfhydryl compounds (Sirchia and Ferrone, 1971) can greatly influence the reactivity of lymphocytes in the cytotoxic test. Therefore, any substance added during the isolation procedure should be removed as completely as possible.

The sensitivity of the inhibition assay is greatly influenced by the alloantisera. Operationally monospecific anti-HL-A alloantisera are required to avoid nonspecific reactions in the inhibition test; if antisera contain a mixture of alloantibodies directed against other known or unknown HL-A and/or non-HL-A specificities, errors may result since those antibodies directed against determinants not present on the soluble HL-A alloantigen under investigation can also react with target cells and thus cause false positive results, as shown in Table V. The sera have to be tested to determine

Table V. Inhibition of the Cytotoxic Reactions of Typing Antisera by Sephadex
Fraction I[a]

HL-A specificity	Phenotype	Absorption	Inhibition
5	+	+	+
5 + X	+	+	−
12	−	−	−
7	+	+	+
1	+	+	+
1 + X	+	−	−
2	+	+	−
3	−	+	+
9	−	−	−

Modified from Kahan et al. (1968).
[a] The phenotype of the donor (N.A.V.) is rather unusual; however, it has been confirmed by
two sets of absorptions. The failure of N.A.V. antigen to inhibit serum 4.16 directed against
HL-A5 + X or serum 16.8 against HL-A1 + X is probably due to the absence of factor X in the
antigenic preparation. Thus, although the antigen may have absorbed the antibody against
HL-A5 or HL-A1, it did not interfere with the action of antibody against factor X.

whether their cytotoxic activity can be specifically inhibited by soluble HL-A
alloantigens, as some alloantisera are inhibited by antigenic preparations
lacking the specificity against which the alloantibodies are directed. Furthermore, some sera routinely and efficiently used in several laboratories for
leukocyte typing may not be suitable for the blocking test, since at any
concentration of the antigenic preparation there is a proportion of target
cells killed. A probable explanation for this phenomenon is the low affinity
of the alloantibodies, resulting in the dissociation of soluble antigen–antibody complexes following the addition of target cells with the antibodies
thus released killing the target cells. HL-A alloantisera must be used at the
highest dilution at which 95% cell death (zero cytotoxicity units) occurs,
because this antibody level has been found most sensitive for detection of
soluble HL-A alloantigen (Pellegrino et al., 1972c, 1973b). In fact, when alloantisera are used at greater concentrations than zero cytotoxic units, larger
quantities of antigens are required to reach the same degree of inhibition
(Fig. 8), causing the test to be less precise. On the other hand, higher dilutions
of antisera produce a lower percentage of dead cells (ranging between 50
and 70%) in the control reaction (alloantiserum without soluble HL-A
alloantigen), and the results are consequently less meaningful and reproducible. For practical reasons, it is advisable to perform inhibition tests at

176   R. A. Reisfeld, S. Ferrone, and M. A. Pellegrino

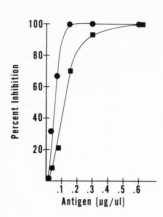

Fig. 8. Influence of the titer of the alloantiserum on the ID_{50} value of soluble HL-A antigen extracted from cultured human lymphoid cells RPMI 1788. The alloantiserum TO-11-03 employed at zero cytotoxic units displays a titer of 1:64 with cell No. 6 (●) and of 1:32 with cell No. 7 (■).

both twofold greater and twofold lower concentrations, since the titer of the alloantiserum can vary from time to time. Selection of sera with high antibody titer is also advisable in order to conserve alloantisera, which are often in short supply. Various HL-A alloantisera directed against the same HL-A specificity, although used at zero cytotoxic units and with the same target cells, can detect a different serological activity of the same soluble HL-A antigen preparation (Kahan *et al.*, 1971; Pellegrino *et al.*, 1972c) (Fig. 9). No relationship seems to exist between the titer of alloantisera and their efficiency in the blocking test (Table VI). The variability in the efficiency of alloantisera in the blocking test may arise from the varying affinities of various alloantibodies for soluble HL-A alloantigens. An alternative mechanism responsible for this phenomenon may be the presence in the operationally monospecific alloantisera of sublytic amounts of additional anti-

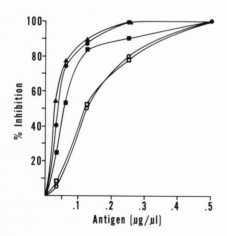

Fig. 9. Variability of ID_{50} values of soluble HL-A antigen extracted from cultured human lymphoid cells RPMI 1788 caused by various anti-HL-A2 antisera employed at zero cytotoxic units with target cell No. 3. ●, TO-11-03; ▲, Eriksson 2-57-0-11-03-01; ■, Stokenberg 1-04-0-07-08-01; ○, Pinquette 2-50-6-09-20-01; □, Stewart.

Table VI. Susceptibilities of Various HL-A Alloantisera to Inhibition by Soluble HL-A Antigens

Anti-HL-A2 antisera	Antigen 115-103 (2,10,7 W14)		Anti-HL-A3 antisera	Antigen 96-103 (3, Te63)		Anti-HL-A5 antisera	Antigen 147-103 (1,2,5, Te57)		Anti-HL-A7 antisera	Antigen 129-103 (2,10,7, W14)	
(name)	Titer[a]	ID_{50}	(name)	Titer	ID_{50}	(name)	Titer	ID_{50}	(name)	Titer	ID_{50}
TO-11-03	32	0.03	Storm	32	0.035	D-66	8	0.13	Cutten	8	0.06
Eriksson	64	0.03	Tucker	128	1.00	McMullen	8	0.13	Melnikoff	8	0.06
Stokenberg	128	0.10	Kraska	4	1.00	Victor	128	0.13	Jackson	32	0.06
Pinquette	128	0.10							Cowen	16	0.13
Stewart	64	0.10							Haas	4	1.14

[a] Expressed as the reciprocal.

bodies directed against different HL-A determinants. Since synergism of HL-A antibodies with different specificities has been shown in the cytotoxic reaction (Ivašková *et al.*, 1969; Svejgaard, 1969; Ahrons and Thorsby, 1970), the amount of alloantibody required to be blocked by the soluble alloantigen in order to inhibit the killing of the target cells will be greater when additional antibodies are present in the reaction mixture. Another possible explanation for the variable efficiency of HL-A alloantisera in the blocking test is the interaction of various components of human sera (aside from that of alloantibodies) with rabbit complement; in fact, human and rabbit complement components are incompatible (N. R. Cooper, unpublished results), and human IgM inhibits natural rabbit antibodies (Herberman, 1970) which play a crucial role in the lymphocytotoxic reaction (Ferrone *et al.*, 1971*b*). The various levels of residual human complement components and the various amounts of IgM present in HL-A alloantisera can influence the lymphocytotoxic test, thus affecting the serological detection of soluble HL-A alloantigens.

Another important source of variability in the inhibition test is the target cells, which can greatly influence the results of the test by a variety of mechanisms. It is known that lymphocytes from different subjects elicit more than one titer level with the same HL-A alloantiserum, and this may

Table VII. Relationship Between the Titer of Cytotoxic HL-A Alloantisera and Sensitivity of the Inhibition Test

Anti-HL-A alloantisera				Antigen ID_{50}		
Name	HL-A specificity	Titer[a]	Target cell No.	115-104 (2,10,7, W14)	96-104 (3, Te63)	147-104 (1,2,5, Te57)
Stokenberg	HL-A2	32	3	3.20	—[b]	—
		64	15	0.24	—	—
		128	1	0.08	—	—
Storm-22-02	HL-A3	16	15	—	0.01	—
		32	1	—	0.03	—
D-66	HL-A5	4	15	—	—	0.75
		8	1	—	—	0.15
Cutten-11-01	HL-A7	8	15	0.26	—	—
		16	1	0.13	—	—

[a] Expressed as the reciprocal.
[b] Not done.

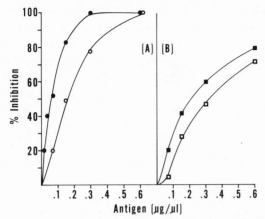

Fig. 10. Effect of various target cells on the ID_{50} value of soluble HL-A antigen extracted from cultured human lymphoid cells RPMI 1788 (HL-A2,10,7, W14). A: Inhibition of TO-11-03 (anti-HL-A2) utilizing target cells No. 1 (●) and No. 13 (○). B: Inhibition of Cutten 2-51-5-01-11-01 (anti-HL-A7) utilizing target cells No. 10 (■) and No. 13 (□).

reflect a variation in the density of HL-A determinants or in the sensitivity of cell membrane to the lytic action of complement. If a given antiserum displays a low titer with a target cell, the amount of antibodies present in the serum at zero cytotoxic units is high; consequently, the quantity of soluble antigen required to block the cytotoxic activity of the HL-A alloantiserum is also comparatively large (Table VII). Thus it is advisable to choose as target cells those lymphocytes against which each respective HL-A allo-antiserum displays the highest possible titer. This maneuver will ensure a high sensitivity of the assay and will allow the conservation of precious alloantisera. Furthermore, the target cells can affect the sensitivity of the inhibition assay by mechanisms other than the titer displayed with alloanti-sera; for instance, in our experiments, (Fig. 10) cells No. 1 and No. 13 have the same titer with alloantiserum TO-11-03 (anti-HL-A2), but the serological activity of the antigenic preparation is markedly increased when cells No. 1 are used as target cells in place of cells No. 13. This variability may reflect the fact that some cells are particularly susceptible to cytolysis because of the greater avidity of the alloantibody for the cell-bound antigen rather than for the soluble antigen. Alternatively, the presence of other HL-A speci-ficities on the target cells can influence the test because of the latent hetero-specificity of the alloantisera or the effect on the reactivity with certain anti-HL-A alloantisera. In this regard, it has been reported that cells presenting HL-A determinants governed by genes in *trans* or *cis* position display a different absorbing capacity (Legrand and Dausset, 1971). Finally, the antigenic determinants against which the natural antibodies of rabbit complement are directed are expressed differently on human peripheral lymphocytes. Thus when a large number of normal rabbit sera were reacted with a panel of unrelated subjects, the lymphocytotoxic reaction was positive only when certain combinations of rabbit sera and human lymphocytes were utilized (Mittal *et al.*, 1973*b*). Cells which best express these antigenic determinants will also detect a higher serological activity of soluble HL-A alloantigens than cells with less expression, because in the latter case the amount of cytotoxic HL-A antibodies to be blocked is considerably greater.

6.2.1. Inhibition Parameters

In order to compare data from different experiments and from different antigen preparations, the following parameters have been utilized and proven to be extremely useful (Kahan *et al.*, 1971; Pellegrino *et al.*, 1972*c*): (1) inhibition dose, (2) avidity coefficient, and (3) specificity ratio.

6.2.1a. Inhibition Dose. The inhibition dose (ID_{50}) represents the amount of soluble HL-A antigen required to halve the cytotoxic activity of operationally monospecific alloantisera; this value arbitrarily represents 1 unit of activity. The reproducibility of the ID_{50} values of antigenic preparations has been assessed by testing an antigenic preparation in triplicate on the same day or

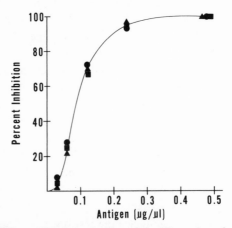

Fig. 11. Reproducibility of the determination of the ID_{50} value of soluble HL-A antigen extracted from cultured human lymphoid cells RPMI 1788 by inhibition of cytotoxic reaction of TO-11-03 (anti-HL-A2) typing alloantiserum, employed at zero cytotoxic units. The reaction was performed in triplicate.

the same antigen preparations on different occasions over a long period of time. The results reported in Fig. 11 and in Table VIII indicate that this parameter is highly reproducible.

Table VIII. Reproducibility of the Inhibition Assay Performed on Different Occasions[a]

Elapsed time (months)	Cell line: HL-A phenotype: Antigen No.: HL-A:	RPMI 1788 2,10,7, W14 11-001		WI-L2 1,2,5, Te57 147-103		RPMI 4098 3, Te63 96-103
		2	7	2	5	3
0		0.018	0.06	0.018	0.075	0.035
$\frac{1}{2}$		0.020	—[b]	—	—	—
1		0.030	—	—	—	—
2		0.021	0.06	0.020	0.055	—
12		0.020	0.07	0.025	0.075	0.045

[a] The results are expressed as ID_{50}.
[b] Not done.

The ID_{50} value indicates (1) presence or absence of any HL-A specificity in the soluble HL-A antigen preparation under investigation, (2) serological activity of the soluble antigen preparation, and (3) degree of purification of the antigenic extract, as less protein will be necessary to achieve a 50% reduction of the cytotoxic activity of the alloantiserum when purification is increased.

6.2.1b. Avidity Coefficient. The avidity coefficient represents the association constant of the soluble HL-A alloantigen–anti-HL-A antibody complex. To calculate the avidity coefficient, the sigmoidal curve which expresses the relationship between the amount of soluble HL-A antigen added to the antiserum and the respective percentage of inhibition of cytotoxicity of anti-HL-A alloantiserum is converted to a linear function by the Reif (1966) modification of the van Krogh equation,

$$\log g = \log G + \frac{1}{m} \log \left(\frac{100 - P}{P} \right)$$

where g is the ratio of total antigen to total antiserum, P is the residual potency (expressed as killed target cell percentage) of the alloantiserum incubated with soluble HL-A alloantigen, G equals g when P is 50%, and

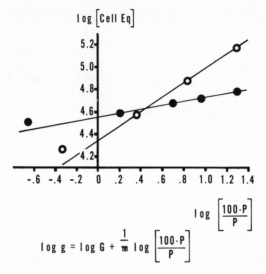

Fig. 12. Van Krogh plot of antigen inhibition. Two allo-antisera recognizing HL-A2, TO-11-03 ($\bigcirc$) and Pinquette ($\bullet$), were preincubated with serial amounts of ammonium sulfate precipitate (expressed as log cell equivalents) of soluble HL-A antigen extracted from cultured lymphoid cells RPMI 1788. The ID_{50} can be seen on the ordinate at the intersection with the linear curve. From Kahan and Reisfeld (1971).

$1/m$ is the slope (Fig. 12). When plotted on a logarithmic scale, the slope of g vs. $(100 - P)/P$ equals $1/m$. In this case, the ID_{50} value can be seen on the ordinate at the intersection with the linear curve. The slope coefficient expresses the avidity of antigen inhibition. Low values of $1/m$ indicate a sharp binding of antibody (i.e., a high AC) with soluble HL-A alloantigen. The ratio between the AC value and the ID_{50} value is arbitrarily expressed as avidity units. The AC is a more sensitive index of the purity of antigenic preparation than the ID_{50} value, as shown in Table IX. In fact, while the ID_{50} units/mg of antigenic preparation increase approximately 25-fold upon purification of the ultracentrifugal supernatant by preparative acrylamide gel electrophoresis, the AC value increases 3200-fold. The increase in the AC value not only reflects an increase in the serological activity of

Table IX. Purification of Human Histocompatibility Antigens

Antigen preparation	ID_{50} units/mg	ID_{50} purification	Avidity units/mg (10^3)	Avidity purification
Ultracentrifugal supernatant	4,000	1.0	70	1
$(NH_4)_2SO_4$ ppt., 30%	10,000	2.5	3,500	50
Acrylamide electrophoresis	100,000	25.0	450,000	3,200

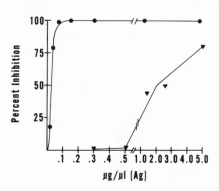

Fig. 13. Specificity ratio. The inhibitory activity of soluble HL-A antigen extracted from cultured human lymphoid cells RPMI 4098 (HL-A3, Te63) against STORM (homologous anti-HL-A3 allo-antiserum, (●) and TO-11-03 (indifferent anti-HL-A2 alloantiserum, ▲) is plotted in arithmetic coordinates.

soluble HL-A antigen as indicated by the higher number of ID_{50} units, but also indicates a sharp absorption of antibody by antigen, signifying a greater avidity for the alloantibody. However, it is worthwhile to emphasize that in order to convert the relationship between the amount of soluble HL-A alloantigen added to the antiserum and the respective percentage of in-hibition of the cytotoxicity of anti-HL-A alloantiserum to a linear function, it is necessary to perform the inhibition assay with arithmetic rather than geometric dilutions of antigenic preparation. Therefore, this procedure is quite time consuming and is not carried out routinely in our laboratory.

6.2.1c. Specificity Ratio. The specificity ratio (SR) (Sanderson, 1968) represents the ratio between the amount of soluble HL-A antigen prepar-ation required for 50 % inhibition of the cytotoxic activity of an alloantiserum (indifferent antiserum) directed against HL-A determinants not detected in the phenotype of cells used for the extraction and the amount required to inhibit an alloantiserum directed against HL-A determinants present in the phenotype of the source of soluble antigen (Fig. 13). While ID_{50} and AC parameters indicate the serological potency of soluble HL-A alloantigens, SR reflects their immunological specificity. The greatest limitation in applying SR is the difficulty in obtaining sufficiently concentrated antigenic prep-arations from which to determine 50 % inhibition of the indifferent serum. At any rate, it has not yet been determined why the cytotoxic activity of indifferent antisera is inhibited by soluble HL-A alloantigens. Although this could be an expression of a nonspecific inhibitory activity caused by non-immunological interactions, other possible explanations can be advanced: (1) At high concentration, the soluble HL-A antigen preparation may acquire anticomplementary activity. Since the lymphocytotoxic test is a complement-dependent reaction, inhibition of indifferent alloantiserum, which is usually reached with large amounts of soluble HL-A alloantigen, could be achieved by a decreased activity of complement. In this regard, it has been observed that soluble HL-A alloantigen preparations can inhibit natural antibodies present in rabbit complement (Mittal *et al.*, 1973*b*). (2) The soluble HL-A alloantigen preparation could interfere with the absorption of alloantibodies on the target cells. It has recently been shown that the lymphocytotoxic reaction is blocked by a variety of nonspecific inhibitors, probably by inhibiting the interaction between antigenic sites and antibodies (Hirata and Terasaki, 1972). (3) Cross-reactivity, well known within the ABO system (Race and Sanger, 1968), is extremely frequent in the HL-A system (Dausset *et al.*, 1968; Svejgaard and Kissmeyer-Nielsen, 1968; Colombani *et al.*, 1970; Thorsby *et al.*, 1970; Ferrone *et al.*, 1972*c*; Mittal and Terasaki, 1972; Mittal *et al.*, 1973*a*; Pellegrino *et al.*, 1973*b*) and could be another explanation. In this case, an indifferent alloantiserum could be inhibited if the soluble HL-A antigen preparation shares determinants in common with those against which the HL-A alloantibodies are directed. (4) Finally, the inhibition of indifferent alloantisera could be an expression of some HL-A determinants within the soluble HL-A antigen preparation which, in turn, are recognized by indifferent alloantisera in the blocking test but not by direct cytotoxicity typing of the cell source (Kahan *et al.*, 1968; Mann and Fahey, 1971).

The value of the SR depends on several variables: (1) the amount of alloantibody employed in the blocking test, (2) homologous and indifferent alloantisera, and (3) the target cells. These factors influence the value of SR by the same mechanisms discussed in Section 6.2.1. Furthermore, it seems worthwhile to emphasize that when choosing the indifferent alloantiserum, it is important to consider the extensive cross-reactivity in the HL-A system. Since small amounts of soluble HL-A alloantigen can inhibit alloantisera directed against determinants cross-reacting with those present on the soluble HL-A antigen (Ferrone *et al.*, 1972*c*), the use of sera with this type of specificity as indifferent antisera can result in an erroneously low value of the SR.

6.2.2. Anticomplementary Factors

As has been mentioned above, the antigen blocking assay utilizes inhibition of the complement-dependent lymphocytotoxic reaction to measure the activity of soluble HL-A antigen. The lymphocytotoxic test does not differ from the lysis of sensitized sheep erythrocytes in the requirement of complement components; all nine complement components are necessary to cause the lysis of sensitized lymphoid cells (Ferrone *et al.*, 1971*b*). In addition, when human lymphocytes were reacted with polyspecific HL-A heteroantisera, a small amount of human complement could accomplish the lymphocytotoxic reaction (A. Ting, S. Ferrone, and P. I. Terasaki, unpublished results). Since the density of HL-A determinants on the surface of peripheral human lymphoid cells is relatively low, the critical concentration of complement at the cell surface necessary to produce a complement–antibody mediated lysis is not reached when cells are reacted with operationally monospecific HL-A alloantisera and human complement components. On the other hand, rabbit serum, which is the most efficient source of complement in the lymphocytotoxic test, contributes to the reaction both complement components and natural antibodies directed against antigenic determinants present on lymphoid cells (Ferrone *et al.*, 1971*b*). Consequently, these antibodies enhance the binding of complement by HL-A antibodies. Any compound used in the antigen blocking test has to be tested for its anti-complementary activity, i.e., possible interference with natural rabbit antibodies and complement components, in order to avoid nonspecific inhibition of the test. The possible effects on complement components can be evaluated by titrating the hemolytic activity of rabbit complement with sensitized sheep erythrocytes in the presence of compounds to be tested. To determine the possible effects on natural rabbit antibodies, a test has been devised which utilizes the same alloantisera and lymphocytes employed as target cells in the inhibition test. This assay is similar to that adopted by Hirata *et al.* (1970) to test the anticomplementary activity of streptococcal M proteins. Lymphocyte suspensions (2000 cells/μl BSS) are incubated in Beckman test tubes (4 by 47 mm) for 1 hr at room temperature with an equal volume of the respective dilutions (0, 1, and 2 cytotoxic units) of the homologous alloantiserum. After incubation, lymphocytes are collected by centrifugation at 13,000 $\times$ g for 5 min in the Beckman microfuge and then readjusted with BSS solution to a final concentration of 2000 cells/μl. One microliter of this suspension is incubated for 30 min at room temperature with 1 μl of the compound under investigation and with 1 μl of BSS. Then rabbit complement (3 μl) is added, and the cytotoxic reaction is performed in the usual way. In a parallel control sample, 1 μl of BSS solution is substituted for the compound under investigation. Another control sample checks the viability of the cells by incubating the sensitized lymphocytes with the compound under investigation and BSS but without complement. Inhibition of the cytotoxic test indicates the presence of anticomplementary activity in the test compound.

6.3. The Quantitative Absorption Test

The absorption test is of great importance in the study of the HL-A system. It is a useful tool to detect false negative reactions (CYNAP) which may occur when HL-A typing is done by the direct cytotoxic test of cells used for the extraction of soluble HL-A antigens. The absorption is useful to detect false positive reactions which may frequently occur in direct HL-A typing of human cultured lymphoid cells; as recently discussed elsewhere (Ferrone *et al.*, 1972*b*), these cells can give positive reactions with those HL-A alloantisera which they are unable to absorb.

Quantitative absorption studies can determine the relative amounts of HL-A determinants present on the surface of lymphoid cells and are thus used to define antigen yields achieved with different extraction and purification procedures.

6.3.1. Procedure

In order to conserve alloantisera, often in short supply, the following microabsorption test is used (Pellegrino *et al.*, 1972*a*). Absorbing cells are washed three times with BSS solution and then suspended at the desired concentration; HL-A alloantisera are used at a dilution corresponding to 1 cytotoxic unit. Five microliters of diluted HL-A alloantisera is incubated with an equal volume of cell suspension at varying concentrations in Beckman microtubes for 60 min at room temperature and mixed every 10 min. At the end of the incubation period, the alloantiserum is cleared by centrifugation at 13,000 × g in a Beckman microfuge for 5 min at 22 °C. The supernatant is transferred to another microtube and tested for residual cytotoxic activity against selected target cells either immediately or after storage at −20 °C. In a parallel control, serum is incubated with an equal volume of BSS solution instead of cell suspension. Different alloantisera with the same HL-A specificity have a different susceptibility to absorption by the same cells; generally, there is a good correlation between susceptibility of cells to absorption of HL-A alloantibody and that to inhibition by soluble HL-A antigen. Therefore, when determining the yield of soluble HL-A alloantigens, it is advisable to employ for the absorption the same alloantisera utilized for the antigen blocking test.

6.3.2. Absorption Parameters

In an attempt to quantitate the expression of HL-A antigens on the cell surface and to determine the number of cells necessary to obtain a given amount of soluble HL-A alloantigen, the following parameters have been found useful (Reisfeld *et al.*, 1971*b*; Pellegrino *et al.*, 1972*c*): (1) absorption dosage, (2) cell equivalent, and (3) percent of recovery.

Absorption dosage (AD_{50}) represents the number of cells required for 50% absorption of the cytotoxic activity of HL-A alloantisera. This parameter is determined by absorbing anti-HL-A alloantisera diluted at 1 cytotoxic unit with twofold progressive dilutions of the cell suspension tested. The percentage of absorption is calculated by the following formula:

$$\% \text{ absorption } = 100 - \frac{\% \text{ cells killed by absorbed serum}}{\% \text{ cells killed by nonabsorbed serum}} \times 100$$

The percentage of absorption of a cytotoxic alloantiserum is plotted on an arithmetic scale against the respective concentration of cells utilized for absorption of the alloantiserum. The relationship between the two parameters is expressed by a sigmoidal curve from which it is possible to calculate the AD_{50} value (Fig. 14).

The AD_{50} parameter indicates (1) presence or absence on the cell surface of any HL-A specificity investigated and (2) relative amount of HL-A determinants per cell. In this way, it is possible to compare the cell surface expression of the same HL-A specificity on different cells as well as the cell surface expression of different HL-A specificities on the same cell. As shown in Table X, the AD_{50} values of several cultured human lymphoid cell lines for the same HL-A specificity show a certain variability, indicating

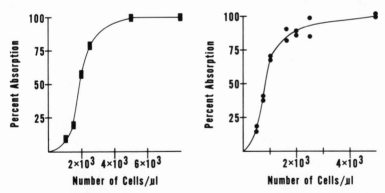

Fig. 14. **Reproducibility of the quantitative microabsorption procedure carried out on two samples of the alloantisera on the same day.** Left: Alloantiserum Cutten 2-51-5-01-11-01 (anti-HL-A7); absorbing cells were cultured human lymphoid cells RPMI 1788 (HL-A2, 10,7, W14). Right: Alloantiserum TO-11-03 (anti-HL-A2); absorbing cells were cultured human lymphoid cells RPMI 1788 (HL-A2,10,7, W14).

Table X. AD_{50}[a] of Human Cultured Lymphoid Cells of Different HL-A Specificities

Cells	HL-A phenotype[b]	First segregant series					Second segregant series			
		HL-A1	HL-A2	HL-A3	HL-A9	HL-A10	HL-A5	HL-A7	HL-A8	W14
RPMI 1788	2,10,7, W14	>50,000	800	>50,000	>15,000	6,000	>50,000	1,800	N.D.	800
RPMI 4980	3, Te63	>50,000	>50,000	600	>50,000	>50,000	>50,000	>50,000	>50,000	>50,000
RPMI 7249	1,2,7,8	500	800	>50,000	—	—	>50,000	3,500	—	—
RPMI 6237	2,3,7	>50,000	1,000	2,000	18,000	>50,000	>50,000	400	>50,000	>50,000
RPMI 8866	2,2,7,12	>50,000	500	1,000	24,000	>50,000	>50,000	400	>50,000	>50,000
WI-L2	1,2,5	800	400	800	3,200	>50,000	1,000	>50,000	>50,000	>50,000
PG-1P-11	2,3,7,12	>50,000	1,600	1,000	>50,000	>50,000	20,000	2,400	>50,000	>50,000
PG-170	3,10,5	>50,000	>50,000	600	>50,000	1,600	1,000	>50,000	>50,000	>50,000
PG-LC-42-F	3,10,5	>50,000	>50,000	300	>50,000	4,000	2,500	>50,000	>50,000	>50,000
PG-1P-7	3,10,5	>50,000	>50,000	600	>50,000	1,600	3,000	>50,000	>50,000	>50,000

[a] Number of cells required to reduce by 50% the cytotoxicity of HL-A alloantisera.
[b] Determined by direct cytotoxicity.

that each cell line has a different density of HL-A antigenic determinants available to react with alloantisera. However, in interpreting these data, some caution should be exercised, since because of the undulating surface and the presence of microvilli it is practically impossible to measure accurately the surface area of cultured human lymphoid cells. Furthermore, the AD_{50} values of a given cell line for homologous HL-A alloantisera with different specificities show a marked variability, suggesting that different HL-A determinants vary quantitatively in their cell surface expression. However, at present it cannot be ruled out that these differences reflect a dissimilar affinity of the HL-A antibodies directed against these various determinants.

6.3.2a. Cell Equivalent. The cell equivalent (CE_{50}) represents the number of cells from which sufficient amounts of soluble HL-A antigen are extracted to achieve a 50 % inhibition of the cytotoxic activity of an HL-A alloantisera. This parameter can be determined by using the following formula:

$$CE = ID_{50} \times \frac{C}{Pr}$$

where C is the number of cells used for the extraction and Pr is the total yield of protein. This parameter is useful when comparing different cell sources used for the extraction of soluble HL-A antigen; for instance, in fetal organs the CE is much lower with kidney cells than with either liver or spleen cells (Table XI) (Pellegrino et al., 1970). In addition, the CE parameter facilitates comparison of different HL-A specificities solubilized from the same cell source (Table XI).

Table XI. Inhibitory Activity and Cell Equivalents

Fetus organ (4 months)	Total cells ($\times 10^9$)	Total protein (μg)	ID_{50}	CE_{50}
Kidney	0.102	1,190	0.06	5,000
Spleen	0.117	245	0.087	40,000
Liver	1.08	506	0.055	135,000
Lung	1.04	5,000	No inhibition	Negative

Modified from Pellegrino et al. (1970).

6.3.2b. Percent Recovery. The percent recovery represents the % ratio between the AD_{50} and the CE_{50} value for any HL-A specificity under investigation. This parameter is determined using the following formula:

$$\% \text{ recovery} = \frac{AD_{50}}{CE_{50}} \times 100$$

It must be stressed that this value is a relative one, since its determination is based only on the amount of those antigenic HL-A determinants which are available to react with alloantisera and not on the total amount of antigen present on, or within, the cells. Although recently it has been reported that HL-A antigens can be detected serologically only on plasma membranes of lymphoid cells (Wilson and Amos, 1972), the possible existence of a pool of histocompatibility antigens inside the cells cannot be ruled out on the basis of presently available data. In addition, it is assumed that the absorption of antibody by the cells is comparable to its blocking by soluble HL-A antigen. However, in the inhibition test, the soluble antigen–antibody complex is still present in the reaction mixture when target cells and complement are added. Moreover, the kinetics of inhibition are different from those of absorption, since in the latter reaction a plateau is reached in less than 1 hr (Fig. 15). In spite of all these limitations, this parameter may be quite useful as a relative value to compare different procedures of antigen extraction and to evaluate the recovery of various antigenic determinants from the same cell source.

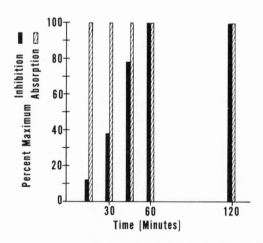

Fig. 15. Comparison of the kinetics of the inhibition and absorption reactions.

6.4. Comments and Conclusions

The serological methods described are simple, rapid, and sensitive means to assess the activity of soluble HL-A antigen preparations; amounts as little as 0.001 μg can readily be detected. The validity of these serological tests is also indicated by the fact that antigenic preparations found inactive by it do not elicit the production of lymphocytotoxic antibodies in rabbits (Pellegrino et al., 1973b). However, it should be pointed out that the results of the lymphocytotoxic test can be influenced by several variables such as (1) time of incubation of soluble HL-A alloantigen with HL-A alloantisera, (2) diluent used for the dilution of antigenic preparation, (3) sequence of incubation of the antigen preparation with target cell and HL-A alloantisera, and (4) different combinations of antisera and target cells. These observations emphasize the need for a standardization of the test system to evaluate the activity of soluble HL-A antigens. Such an approach will make it feasible to compare in a meaningful way results from different laboratories utilizing various methods of extraction and purification of soluble HL-A alloantigens derived from a variety of cell sources.

However, the above discussion points out that the inhibition test is a complex and, as yet, not completely understood reaction which can be influenced by a variety of compounds and experimental conditions. Therefore, much caution should be exercised in drawing conclusions from the results of these tests. It seems to us hazardous to infer conclusions on the chemical composition of histocompatibility antigens from the ability of chemically pure compounds, e.g., synthetic carbohydrates (Davies, 1966), to inhibit the lymphocytotoxic reaction. Finally, the limitations of these serological tests should be emphasized, since these assays evaluate only an in vitro reactivity of cytotoxic HL-A alloantisera with these antigens without shedding any light on their biological function and immunogenicity. Therefore, the results of these in vitro tests do not establish whether or not soluble HL-A alloantigens are true transplantation antigens, i.e., substances which specifically affect allograft survival. However, at present the serological tests are invaluable in purification and structure studies of soluble HL-A antigens since they are speedy, precise, and specific.

7. PERSPECTIVES

H-antigens as genetically determined individuality markers on mammalian cell surfaces present unique opportunities for the study of transplantation individuality, cell recognition, and contact phenomena. Advances in the serological evaluation of HL-A antigens permit the mapping of these markers on the cell surface of cultured lymphoid cells and the quantitative study of their expression during the cell growth cycle. Progress in the elucidation of the physicochemical properties of HL-A antigens now allows the construction of a molecular model in which differences in amino acid composition define the antigenic specificity in regions of allotypic H-markers. During the elucidation of the biological and chemical characteristics of the HL-A system, which offers excellent cell surface markers, other allotypic systems might be uncovered, thereby contributing to a better understanding of the molecular structure and functional properties of the mammalian cell surface membrane.

ACKNOWLEDGMENTS

This is publication No. 629 from the Department of Experimental Pathology, Scripps Clinic and Research Foundation, La Jolla, California. This work was supported by United States Public Health Service grants AI 10180 and CA 10596 from the National Institutes of Health, National Cancer Institute contract 72-2046, and grant IC-5 from the American Cancer Society. Dr. Ferrone is a visiting scientist from the University of Milan, Italy.

8. REFERENCES

Ahrons, S., and Thorsby, E., 1970, Cytotoxic HL-A antibodies. Studies of synergism and gene–dose effect, *Vox Sang.* **18**:323.

Amos, D. B., 1953, The agglutination of mouse leucocytes by iso-immune sera, *Brit. J. Exptl. Pathol.* **34**:464.

Amos, D. B., 1966, Some results on the cytotoxicity test, *in* "Histocompatibility Testing" (D. B. Amos and J. J. van Rood, eds.), pp. 151–157, Williams and Wilkins, Baltimore.

Aoki, T., Hämmerling, U., de Harven, E., Boyse, E. A., and Old, L. J., 1969, Antigenic structure of cell surface. An immunoferritin study of the occurrence and topography of H-2, Θ, and TL alloantigens on mouse cells, *J. Exptl. Med.* **130**:979.

Aoki, T., Boyse, E. A., Old, L. J., de Harven, E., Hämmerling, U., and Wood, H. A., 1970, G (Gross) and H-2 cell surface antigens: Location on Gross leukemia cells by electron microscopy with visually labeled antibody, *Proc. Natl. Acad. Sci. (U.S.)* **65**:569.

Blandamer, A. H., O'Neill, J., and Summerell, J. M., 1969, *Widerbelebung Organ. Intensivmed.* **6**:62.

Boyse, E. A., Old, L. J., and Stockert, E., 1968, An approach to the mapping of antigens on the cell surface, *Proc. Natl. Acad. Sci. (U.S.)* **60**:886.

Braun, W. E., Grecek, D. R., and Murphy, J. J., 1972, Expanded HL-A phenotypes of human peripheral lymphocytes after trypsinization, *Transplantation* **13**:337.

Burnet, F. M., 1970, A certain symmetry: Histocompatibility antigens compared with immunocyte receptors, *Nature (Lond.)* **226**:123.

Burton, K., 1956, A study of conditions and mechanisms of the diphenylamine reaction for the colorimetric estimation of deoxyribonucleic acid, *Biochem. J.* **62**:315.

Cerottini, J. C., and Brunner, K. T., 1967, Localization of mouse isoantigens on the cell surface as revealed by immunofluorescence, *Immunology* **13**:395.

Chapel, H. M., and Welsh, K. I., 1972, Removal and resynthesis of HL-A2, 5 and 12 determinants at the surface of peripheral blood lymphocytes, *Transplantation* **13**:347.

Colombani, J., Colombani, M., and Dausset, J., 1970, Cross-reactions on the HL-A system with special reference to Da6 cross-reacting group. Description of HL-A antigens Da 22, Da 23, Da 24 defined by platelet complement fixation, *in* "Histocompatibility Testing 1970" (P. I. Terasaki, ed.), pp. 79–92, Munksgaard, Copenhagen.

Dandliker, W. B., and de Saussure, V. A., 1971, Stabilization of macromolecules by hydrophobic bonding: Role of water structure and of chaotropic ions, *in* "The Chemistry of Biosurfaces" (M. L. Hair, ed.), pp. 1–40, M. Dekker, New York.

Dandliker, W. B., Alonso, R., de Saussure, V. A., Kierszenbaum, F., Levison, S. A., and Shapiro, H. C., 1967, The effect of chaotropic ions on the dissociation of antigen–antibody complexes, *Biochemistry* **6**:1460.

Danielli, J. F., and Davson, H., 1935, A contribution to the theory of permeability of thin films, *J. Cell. Comp. Physiol.* **5**:495.

Dausset, J., Colombani, J., Colombani, M., Legrand, L., and Feingold, N., 1968, Un nouvel antigène du système HL-A (Hu-1): l'Antigène 15, allèle possible des anti-gènes 1, 11, 12, *Nouv. Rev. Fr. Hématol.* **8**:398.

Dausset, J., Colombani, J., Legrand, L., and Fellous, M., 1970, Genetics of the HL-A system: Deduction of 480 haplotypes, *in* "Histocompatibility Testing 1970" (P. I. Terasaki, ed.), pp. 53–75, Munksgaard, Copenhagen.

Davies, D. A. L., 1966, Histocompatibility antigens of the mouse, Immunopathol. 4th Internat. Symp. 1965, p. 111.

Davies, D. A. L., 1969, The molecular individuality of different mouse H-2 histocompatibility specificities determined by single genotypes, *Transplantation* **8**:51.

Davies, D. A. L., Manstone, A. J., Viza, D. C., Colombani, J., and Dausset, J., 1968, Human transplantation antigens: The HL-A (Hu-1) system and its homology with the mouse H-2 system, *Transplantation* **6**:571.

Davis, W. C., 1972, H-2 antigen on cell membranes: An explanation for the alteration of distribution by indirect labeling techniques, *Science* **175**:1006.

Davis, W. C., and Silverman, L., 1968, Localization of mouse H-2 histocompatibility antigen with ferritin-labeled antibody, *Transplantation* **6**:535.

Davis, W. C., Alspaugh, M. A., Stimpfling, J. H., and Walford, R. L., 1971, Cellular surface distribution of transplantation antigens: Discrepancy between direct and indirect labeling techniques, *Tissue Antigens* **1**:89.

Engelfriet, C. P., and Britten, A., 1965, The cytotoxic test for leucocyte antibodies: A simple and reliable technique, *Vox Sang.* **10**:660.

Etheredge, E. E., and Najarian, J. S., 1971, Solubilization of human histocompatibility substances, *Transplant. Proc.* **3**:224.

Ferrone, S., Tosi, R. M., and Centis, D., 1967, Anticomplementary factors affecting the lymphocytotoxicity test, *in* "Histocompatibility Testing 1967" (E. S. Curtoni, P. L. Mattiuz, and R. M. Tosi, eds.), pp. 375–382, Munksgaard, Copenhagen.

Ferrone, S., Pellegrino, M. A., and Reisfeld, R. A., 1971a, A rapid method for direct HL-A typing of cultured lymphoid cells, *J. Immunol.* **107**:613.

Ferrone, S., Cooper, N. R., Pellegrino, M. A., and Reisfeld, R. A., 1971b, The lymphocytotoxic reaction: The mechanism of rabbit complement action, *J. Immunol.* **107**:939.

Ferrone, S., Natali, P. G., Hunter, A., Terasaki, P. I., and Reisfeld, R. A., 1972a, Immunogenicity of soluble HL-A alloantigens, *J. Immunol.* **108**:1718.

Ferrone, S., Pellegrino, M. A., and Reisfeld, R. A., 1972b, Typing of cultured human lymphoid cells, *Lancet* **1**:1237.

Ferrone, S., Mittal, K. K., Pellegrino, M. A., Terasaki, P. I., and Reisfeld, R. A., 1972c, Serological specificity and crossreactivity of soluble HL-A alloantigens, *Immunol. Comm.*, **1**:71.

Ferrone, S., Pellegrino, M. A., Götze, D., Mittal, K. K., Terasaki, P. I., and Reisfeld, R. A., 1973a, Cytotoxic heteroantisera against soluble HL-A antigens, *in* "Proceedings of the International Symposium on Standardization of HL-A Reagents, Copenhagen," Vol. 18, p. 218, Karger, Basel/New York.

Ferrone, S., Cooper, N. R., Pellegrino, M. A., and Reisfeld, R. A., 1973b, Interaction of histocompatibility (HL-A) antibodies and complement with synchronized human lymphoid cells in continuous culture, *J. Exp. Med.* **137**:55.

Frye, L. D., and Edidin, M., 1970, The rapid intermixing of cell surface antigens after formation of mouse–human heterokáryons, *J. Cell Sci.* **7**:319.

Gervais, A. G., 1968, Detection of mouse histocompatibility antigens by immunofluorescence, *Transplantation* **6**:261.

Gibofsky, A., and Terasaki, P. I., 1972, Trypsinization of lymphocytes for HL-A typing, *Transplantation* **13**:192.

Gorer, P. A., and Mikulska, Z. B., 1954, The antibody response to tumor inoculation: Improved methods of antibody detection, *Cancer Res.* **14**:651.

Götze, D., and Reisfeld, R. A., 1972, Evaluation of immunogenic and tolerogenic properties of soluble H-2 antigen, *Fed. Proc.* **31**:2419.

Greenbaum, L. M., 1971, Cathepsins and kinin-forming and destroying enzymes, *in* "The Enzymes" (P. D. Boyer, ed.), p. 475, Academic Press, New York.

Grothaus, E. A., Wayne, F. M., Yunis, E., and Amos, D. B., 1971, Human lymphocyte antigen reactivity modified by neuraminidase, *Science* **173**:542.

Hamaguchi, K., and Geiduschek, E. P., 1962, Effect of electrolytes on the stability of the deoxyribonucleate helix, *J. Am. Chem. Soc.* **84**:1329.

Hatefi, Y., and Hanstein, W. G., 1968, Solubilization of particulate proteins and nonelectrolytes by chaotropic agents, *Proc. Natl. Acad. Sci. (U.S.)* **62**:1129.

Haughton, G., 1964, Extraction of H-2 antigen from mouse tumor cells, *Transplantation* **2**:251.

Haughton, G., 1966, Transplantation antigen of mice: Cellular localization of antigen determined by the H-2 locus, *Transplantation* **4**:238.

Herberman, R. B., 1970, Inhibition of natural cytotoxic rabbit antibody by human IgM: Production of non toxic rabbit serum for use as complement source, *J. Immunol.* **104**:805.

Hirata, A. A., and Terasaki, P. I., 1972, Masking of human transplantation antigens by diverse substances, *J. Immunol.* **108**:1542.

Hirata, A. A., Armstrong, A. S., Kay, J. W. D., and Terasaki, P. I., 1970, Specificity of inhibition of HL-A antisera by streptococcal M proteins, *in* "Histocompatibility Testing 1970" (P. I. Terasaki, ed.), pp. 475–481, Munksgaard, Copenhagen.

Hoelzl-Wallach, D. F. H., and Gordon, A. S., 1969, The structure of cellular membranes, *in* "Cellular Recognition" (R. T. Smith and R. A. Good, eds.), p. 3, Appleton, New York.

Hofmeister, F., 1888, *Arch. Exptl. Pathol. Pharmakol.* **24**:247.

Ivaščová, E., Vybiralová, H., Raue, I., Démant, P., and Ivanyi, P., 1969, Synergic action of HL-A antibodies, *Folia Biol. (Praha)* **15**:26.

Kahan, B. D., 1965, Isolation of a soluble transplantation antigen, *Proc. Natl. Acad. Sci. (U.S.)* **53**:153.

Kahan, B. D., and Reisfeld, R. A., 1968, Differences in the amino acid composition of allogeneic guinea pig transplantation antigens, *J. Immunol.* **101**:237.

Kahan, B. D., and Reisfeld, R. A., 1969a, Transplantation antigens, *Science* **164**:514.

Kahan, B. D., and Reisfeld, R. A., 1969b, Advances in the chemistry of transplantation antigens, *Transplant. Proc.* **1**:483.

Kahan, B. D., and Reisfeld, R. A., 1971, Chemical markers of transplantation individuality solubilized with sonic energy, *Bacteriol. Rev.* **35**:59.

Kahan, B. D., Reisfeld, R. A., Pellegrino, M., Curtoni, E. S., Mattiuz, P. L., and Ceppellini, R., 1968, Water-soluble human transplantation antigens, *Proc. Natl. Acad. Sci. (U.S.)* **61**:897.

Kahan, B. D., Pellegrino, M. A., Papermaster, B. W., and Reisfeld, R. A., 1971, Quantitative serologic parameters of purified HL-A antigens, *Transplant. Proc.* **3**:227.

Kandutsch, A. A., 1960, Intracellular distribution and extraction of tumor homograft-enhancing antigens, *Cancer Res.* **20**:262.

Kandutsch, A. A., 1963, Partial purification of tissue isoantigens from a mouse sarcoma, *Transplantation* **1**:201.

Kauzman, W., 1959, Some factors in the interpretation of protein denaturation, *Advan. Protein Chem.* **14**:1.

Kissmeyer-Nielsen, F., and Thorsby, E., 1970, Human transplantation antigens, *Transplant. Rev.* **4**:1.

Klotz, I. M., and Farnham, S. B., 1968, Stability of an amide–hydrogen bond in an apolar environment, *Biochemistry* **7**:3879.

Kourilsky, F. M., Silvestre, D., Levy, J. P., Dausset, J., Nicolai, M. G., and Senik, A. J., 1971, Immunoferritin study of the distribution of HL-A antigens of human blood cells, *J. Immunol.* **106**:454.

Legrand, L., and Dausset, J., 1971, Cell surface interaction between HL-A antigen determinants, *Nature New Biol.* **234**:271.

Lunney, J., Chrambach, A., and Rodbard, D., 1971, Factors affecting resolution, band width, number of theoretical plates, and apparent diffusion coefficients in poly-acrylamide gel electrophoresis, *Anal. Biochem.* **40**:158.

Mann, D. L., 1972, Comparisons of HL-A alloantigens solubilized by papain and TIS, *in* "Chemical Markers of Biological Individuality: The Transplantation Antigens" (B. D. Kahan and R. A. Reisfeld, eds.), pp. 287, Academic Press, New York.

Mann, D. L., and Fahey, J. L., 1971, Histocompatibility antigens, *Ann. Rev. Microbiol.* **25**:679.

Mann, D. L., and Levy, R., 1971, Solubilization of HL-A alloantigens with SDS, *Fed. Proc.* **30**:2767.

Mann, D. L., Rogentine, G. N., Fahey, J. L., and Nathenson, S. G., 1968, Solubilization of human leukocyte membrane isoantigens, *Nature (Lond.)* **217**:1180.

Mann, D. L., Rogentine, G. N., Fahey, J. L., and Nathenson, S. G., 1969, Human lymphocyte membrane (HL-A) alloantigens: Isolation, purification and properties, *J. Immunol.* **103**:282.

Manson, L. A., and Palm, J., 1972, Intracellular distribution of transplantation antigens, *in* "Chemical Markers of Biological Individuality: The Transplantation Antigens" (B. D. Kahan and R. A. Reisfeld, eds.), p. 141, Academic Press, New York.

Mason, S., and Warner, N. L., 1970, The immunoglobulin nature of the antigen recognition site on cells mediating transplantation immunity and delayed hypersensitivity, *J. Immunol.* **104**:762.

McDonald, J. C., Jacobbi, L., and Williams, R. W., 1970, HL-A antigen studies with leukocyte cell lines, *Transplantation* **10**:499.

Meltzer, M. S., Leonard, E. J., Rapp, H. J., and Borsos, T., 1971, Tumor-specific antigen solubilized by hypertonic potassium chloride, *J. Natl. Cancer Inst.* **47**:703.

Mercuriali, F., Richiardi, P., Mattiuz, P. L., and Sirchia, G., 1971, AET-treated lympho-cytes; their use in the lymphocytotoxicity reaction, *Tissue Antigens* **1**:290.

Mittal, K. K., and Terasaki, P. I., 1972, Cross-reactivity in the HL-A system, *Tissue Antigens* **2**:94.

Mittal, K. K., Mickey, M. R., Singal, D. P., and Terasaki, P. I., 1968, Serotyping for homotransplantation. XVIII. Refinement of microdroplet lymphocyte cytotoxicity test, *Transplantation* **6**:913.

Mittal, K. K., Mickey, M. R., and Terasaki, P. I., 1973a, Cross-reactive alloantibodies and alloantigens of the HL-A system, *in* "Proceedings of the International Sym-posium on Standardization of HL-A Reagents, Copenhagen," Vol. 18, p. 165, Karger, Basel/New York.

Mittal, K. K., Ferrone, S., Reisfeld, R. A., Mickey, M. R., Pellegrino, M. A., and Terasaki, P. I., 1973b, Specificity of lymphocytotoxic anti-human antibodies in normal rabbit sera, *Tissue Antigens*, **3**:88.

Möller, G., 1961, Demonstration of mouse isoantigens at the cellular level by the fluorescent antibody technique, *J. Exptl. Med.* **114**:415.

Moore, G. E., Gerner, R. F., and Franklin, H. A., 1967, Culture of normal human leukocytes, *J. Am. Med. Ass.* **199**:519.

Muramatsu, T., and Nathenson, S. G., 1971, Carbohydrate structure of mouse H-2 alloantigens, *Fed. Proc.* **30**:2768.

Nelson, R. A., Jr., and Biro, C. E., 1968, Complement components of a haemolytically deficient strain of rabbits, *Immunology* **14**:527.

Nicolson, G. L., Masouredis, S. P., and Singer, J., 1971, Quantitative two-dimensional ultrastructural distribution of Rh_0 (D) antigenic sites on human erythrocyte membranes, *Proc. Natl. Acad. Sci. (U.S.)* **68**:1416.

Ornstein, L., 1964, Disc electrophoresis in polyacrylamide gels, *Ann. N.Y. Acad. Sci.* **121**:321.

Ozer, J. H., and Hoelzl-Wallach, D. F., 1967, H-2 components and cellular membranes, *Transplantation* **5**:652.

Pellegrino, M. A., Pellegrino, A., and Kahan, B. D., 1970, Solubilization of fetal HL-A antigens, *Transplantation* **19**:425.

Pellegrino, M. A., Ferrone, S., and Pellegrino, A., 1972a, A simple microabsorption technique for HL-A typing, *Proc. Soc. Exptl. Biol. Med.* **139**:484.

Pellegrino, M. A., Ferrone, S., Natali, P. G., Pellegrino, A., and Reisfeld, R. A., 1972b, Expression of HL-A antigens in synchronized cultures of human lymphocytes, *J. Immunol.* **108**:573.

Pellegrino, M. A., Ferrone, S., Pellegrino, A., and Reisfeld, R. A., 1973a, The expression of HL-A antigens during the growth cycle of cultured human lymphoid cells, *Clin. Immunol. Immunopathol.*, **1**:182.

Pellegrino, M. A., Ferrone, S., and Pellegrino, A., 1972c, Serological detection of soluble HL-A antigens, *in* "Chemical Markers of Biological Individuality: The Transplantation Antigens" (B. D. Kahan and R. A. Reisfeld, eds.), p. 433, Academic Press, New York.

Pellegrino, M. A., Ferrone, S., Pellegrino, A., and Reisfeld, R. A., 1973b, A critical evaluation of the serologic assay for soluble HL-A alloantigens, *in* "Proceedings of the International Symposium on Standardization of HL-A Reagents, Copenhagen," Vol. 18, p. 209 (Karger, Basel/New York).

Pellegrino, M. A., Ferrone, S., Mittal, K. K., Pellegrino, A., and Reisfeld, R. A., 1973c, A quantitative study of cross-reactivity in the HL-A system with human cultured lymphoid cells and soluble HL-A antigens, *Transplantation*, **15**:42.

Race, R. R., and Sanger, E., 1968, *in* "Blood Groups in Man," 5th ed., Blackwell, Oxford.

Rapaport, F. T., Dausset, J., Converse, J. M., and Lawrence, H. S., 1965, Biologic and ultrastructural studies of leukocyte fractions as transplantation antigens in man, *Transplantation* **3**:490.

Raymond, S., and Weintraub, L., 1959, Acrylamide gel as a supporting medium for zone electrophoresis, *Science* **130**:711.

Reif, A. E., 1966, An experimental test of two general relationships to describe the absorption of antibodies by cells and tissue, *Immunochemistry* **3**:267.

Reisfeld, R. A., 1967, Heterogeneity of rabbit light-polypeptide chains, *Proc. Cold Spring Harbor Symp. Quant. Biol.* **32**:291.

Reisfeld, R. A., and Kahan, B. D., 1970a, Transplantation antigens, *Advan. Immunol.* **12**:117.

Reisfeld, R. A., and Kahan, B. D., 1970b, Biological and chemical characterization of human histocompatibility antigens, *Fed. Proc.* **29**:2034.

Reisfeld, R. A., and Kahan, B. D., 1971, Extraction and purification of soluble histocompatibility antigens, *Transplant. Rev.* **6**:81.

Reisfeld, R. A., and Kahan, B. D., 1972a, Human histocompatibility antigens, in "Contemporary Topics in Immunochemistry" (F. P. Inman, ed.), pp. 51–92, Plenum Press, New York.

Reisfeld, R. A., and Kahan, B. D., 1972b, The molecular nature of HL-A antigens, in "Chemical Markers of Biological Individuality: The Transplantation Antigens" (B. D. Kahan and R. A. Reisfeld, eds.), p. 489, Academic Press, New York.

Reisfeld, R. A., and Pellegrino, M. A., 1972, Salt extraction of soluble HL-A antigens, in "Chemical Markers of Biological Individuality: The Transplantation Antigens" (B. D. Kahan and R. A. Reisfeld, eds.), p. 259, Academic Press, New York.

Reisfeld, R. A., Lewis, U. J., Brink, N. G., and Steelman, J. L., 1962, Purification of human growth hormone, *Endocrinology* **71**:559.

Reisfeld, R. A., Börjeson, J., Chessin, L. N., and Small, P. A., 1967, Isolation and characterization of a mitogen from pokeweed *(Phytolacca americana)*, *Proc. Natl. Acad. Sci. (U.S.)* **58**:2020.

Reisfeld, R. A., Pellegrino, M., Papermaster, B. W., and Kahan, B. D., 1970, HL-A antigens from a continuous lymphoid cell line derived from a normal donor. I. Solubilization and serologic characterization, *J. Immunol.* **104**:560.

Reisfeld, R. A., Pellegrino, M. A., and Kahan, B. D., 1971a, Salt extraction of soluble HL-A antigens, *Science* **172**:1134.

Reisfeld, R. A., Pellegrino, M., Papermaster, B. W., and Kahan, B. D., 1971b, Progress in the isolation of HL-A antigens, in "Immunopathology: Sixth International Symposium" (P. A. Miescher, ed.), p. 139, Grune and Stratton, New York.

Rodbard, D., and Chrambach, A., 1971, Estimation of molecular radius, free mobility and valence using polyacrylamide gel electrophoresis, *Anal. Biochem.* **40**:95.

Rodbard, D., Kapadia, G., and Chrambach, A., 1971, Pore gradient electrophoresis, *Anal. Biochem.* **40**:135.

Rogentine, G. N., Jr., and Gerber, P., 1969, HL-A antigens of human lymphoid cells in long-term tissue cultures, *Transplantation* **8**:28.

Rogentine, G. N., Jr., and Gerber, P., 1970, Qualitative and quantitative comparisons of HL-A antigens on different lymphoid cell types from the same individuals, in "Histocompatibility Testing 1970" (P. I. Terasaki, ed.), pp. 333–338, Munksgaard, Copenhagen.

Sanderson, A. R., 1968, HL-A substances from human spleens, *Nature (Lond.)* **220**:192.

Sanderson, A. R., and Batchelor, J. R., 1968, Transplantation antigens from human spleens, *Nature (Lond.)* **219**:184.

Sanderson, A. R., and Welsh, K. I., 1972a, Purification and structural studies of alloantigen determinants solubilized by papain, in "Chemical Markers of Biological Individuality: The Transplantation Antigens" (B. D. Kahan and R. A. Reisfeld, eds.), p. 273, Academic Press, New York.

Sanderson, A. R., and Welsh, K. I., 1972b, The study of isoimmune antibodies and antigenic determinants using the [51]chromium cytotoxic assay, in "Chemical Markers of Biological Individuality: The Transplantation Antigens" (B. D. Kahan and R. A. Reisfeld, eds.), p. 453, Academic Press, New York.

Sanderson, A. R., Cresswell, P., and Welsh, K. I., 1971, Involvement of carbohydrate in the immunochemical determinant area of HL-A substances, *Nature New Biol.* **230**:8.

Shimada A., and Nathenson, S. G., 1969, Murine histocompatibility H-2 alloantigens, *Biochemistry* **8**:4048.

Singer, S. J., and Nicolson, G. L., 1972, The fluid mosaic model of the structure of cell membranes, *Science* **175**:720.

Sirchia, G., and Ferrone, S., 1971, Normal human lymphocyte treated *in vitro* with the sulfhydril compound AET: Relationship to the lymphocytes of PNH, *Blood* **37**:563.

Svejgaard, A., 1969, Synergistic action of HL-A isoantibodies, *Nature (Lond.)* **222**:94.

Svejgaard, A., and Kissmeyer-Nielsen, F., 1968, Cross-reactive human HL-A isoantibodies, *Nature (Lond.)* **219**:868.

Takasugi, M., 1971, An improved fluorochromatic cytotoxic test, *Transplantation* **12**:148.

Terasaki, P., and McClelland, J., 1964, Microdroplet assay of human serum cytotoxins, *Nature (Lond.)* **204**:998.

Terasaki, P. I., Vredevoe, D. L., and Mickey, M. R., 1967, Serotyping for homotransplantation. X. Survival of 196 grafted kidneys subsequent to typing, *Transplantation* **5**:1057.

Thomas, L., 1959, Mechanisms involved in tissue damage by endotoxins of gram-negative bacteria, *in* "Cellular and Humoral Aspects of the Hypersensitive State," pp. 451–468, Wiley and Sons, London.

Thorsby, E., Kjerbye, K. E., and Bratlie, A., 1970, Cross-reactive HL-A antibodies: Absorption on immunization studies, *Vox Sang.* **18**:373.

Uhlenbruck, G., Voigtmann, R., Salfner, B., Bube, F. W., and Seibel, E., 1973, Comparison of different methods for the solubilization of HL-A substances, *in* "Proceedings of the International Symposium on Standardization of HL-A Reagents, Copenhagen," Vol. 18, p. 205 (Karger, Basel/New York).

Walford, R. L., Gallagher, R., and Sjaarda, J. R., 1964, Serologic typing of human lymphocytes with immune serum obtained after homografting, *Science* **144**:868.

Walford, R. L., Shanbrom, E., Troup, G. M., Zeller, E., and Ackerman, B., 1967, Lymphocyte grouping with defined antisera, *in* "Histocompatibility Testing 1967" (E. S. Curtoni, P. L. Mattiuz, and R. M. Tosi, eds.), p. 221, Munksgaard, Copenhagen.

Warren, L., and Glick, M. C., 1968, Membranes of animal cells. II. The metabolism and turnover of the surface membrane, *J. Cell Biol.* **37**:729.

Warren, L., Glick, M. C., and Nass, M. K., 1966, Membranes of animal cells. I. Methods of isolation of the surface membrane, *J. Cell. Physiol.* **68**:269.

Willingham, M. C., Spicer, S. S., and Graber, C. D., 1971, Immunocytologic labeling of calf and human lymphocyte surface antigens, *Lab. Invest.* **25**:211.

Wilson, L. A., and Amos, D. B., 1972, Subcellular location of HL-A antigens, *Tissue Antigens* **2**:105.

Yunis, E. J., Ward, F. E., and Amos, D. B., 1970, Observations of the CNAP phenomenon, *in* "Histocompatibility Testing 1970" (P. I. Terasaki, ed.), p. 351, Munksgaard, Copenhagen.

zur Hausen, H., Diehl, V., Wolf, H., Schulte-Holthausen, H., and Schneider, U., 1972, Occurrence of Epstein–Barr virus genomes in human lymphoblastoid cell lines, *Nature New Biol.* **237**:189.

Chapter 5

Dissociation and Reassembly of the Inner Mitochondrial Membrane

YASUO KAGAWA

Department of Biochemistry
Jichi Medical School
Kawachi-gun, Tochigi-ken, Japan

1. COMPONENTS OF THE INNER MITOCHONDRIAL MEMBRANE

1.1. Introduction

1.1.1. Membrane Proteins and Phospholipids

Oxidative phosphorylation is the synthesis of ATP by energy liberated during substrate oxidation (Lardy and Ferguson, 1969; Slater, 1971). This function is localized in the inner membrane of mitochondria (Fig. 1A), which is mainly composed of many kinds of proteins and several phospholipids. Because of this complexity, dissociation, fractionation, and reassembly of the membrane components are necessary to understand oxidative phosphorylation (Racker, 1970; Kagawa, 1972*a*).

Functionally, these proteins can be divided into three groups:

1. Electron transport system, such as flavoproteins and cytochromes (Keilin, 1930; Yakushiji and Okunuki, 1940; Hatefi, 1966).
2. Energy transfer system, including mitochondrial ATPase [ATP phosphohydrolase, E.C. 3.6.1.3] (Pullman *et al.*, 1960) (Fig. 1G,H) and oligomycin-sensitivity-conferring proteins (Kagawa and Racker, 1966*a*; MacLennan and Tzagoloff, 1968).
3. Other proteins, such as carrier proteins of ions and nucleotides.

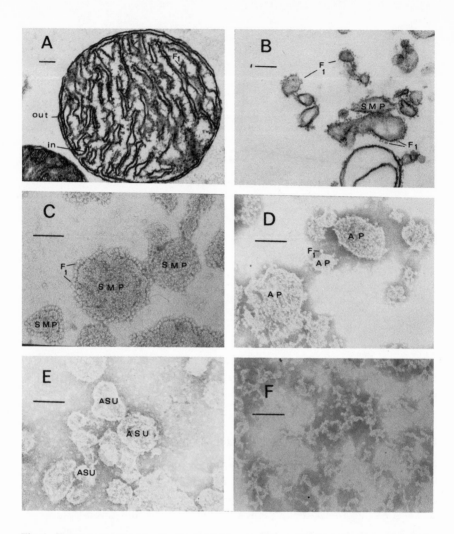

Fig. 1. **Electron micrographs of mitochondrial preparations.** The samples were negatively stained (C–I and K) as described in Kagawa and Racker (1966c) or thin-sectioned and positively stained (A, B, J, and L) as described in Kagawa and Racker (1971) with a modification by Dr. J. Telford. The magnification scale of 1000 Å is indicated as the bar in the upper left of each photograph (A and L, ×60,000; B, ×80,000; C–G and I–K, ×120,000; H, ×600,000). The symbol F_1 indicates F_1 (ATPase) molecules. A: Intact beef-heart mitochondria. out, Outer membrane; in, inner membrane. B: Submitochondrial particles (SMP) prepared as described in Section 6.1.4. C: Same as B. D: Alkali-treated particles (AP) as described in Section 6.1.2. E: Alkali–Sephadex–urea treated particles (ASU) as described in Section 6.1.2. F: Complex I (NADH-CoQ reductase) described by Hatefi (1966). G: Purified F_1 as described in Section 5.1.1. H: Purified F_1, partially crystallized. I: Oligomycin-sensitive ATPase. J: Dialyzed phospholipids. K: Vesicles reconstituted as described in Section 7.2.3. L: Same as K.

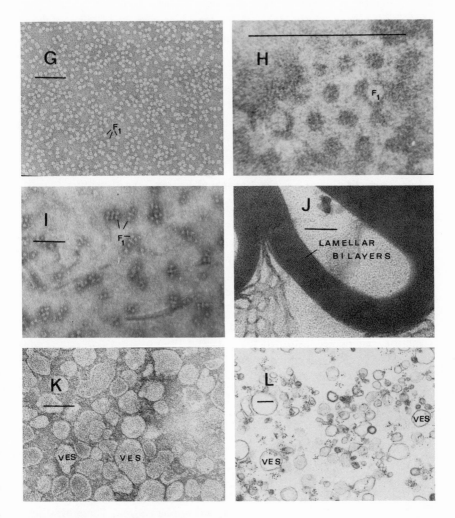

1.1.2. Analogy of Multienzyme Systems

Fractionation and reconstitution of multienzyme systems have been successful in the case of soluble systems such as the glycolytic pathway. However,
contrary to the soluble systems where a substrate is degraded by a series of
enzymes connected with intermediary metabolic compounds step by step,
the oxidative phosphorylation system has no corresponding compounds.
In this kind of insoluble membrane-bound system, flow of electrons through
the electron transport system, flow of energy by either electrochemical
potential of ions (Cockrell *et al.*, 1967), or conformational change of proteins

(Ryrie and Jagendorf, 1972) may be the major reaction. The lipid-containing vesicular membrane structure in which the abovementioned components are arranged in an anisotropic way may be essential in oxidative phosphorylation (Mitchell, 1966). This structural organization is one of the difficult points in the reassembly process of this system.

The regulatory mechanism of oxidative phosphorylation is called "respiratory control." A soluble multienzyme system is regulated by allosteric effects, while respiratory control may be caused by the electrochemical potential of ions distributed inside and outside of the vesicles.

1.1.3. Reconstitution of Vesicles Capable of Energy Transformation

It is possible to extend the theoretical analogy of soluble systems with their chemical intermediates (Slater, 1966, 1971) to oxidative phosphorylation, but phosphorylation in the absence of vesicular membrane has, to date, not been achieved. However, it has been possible to reconstitute vesicles capable of energy transformation (ATP synthesis, H^+ accumulation, membrane potential formation, and ^{32}Pi-ATP exchange) from solubilized membrane components (Kagawa and Racker, 1971; Racker and Kandrach, 1971; Kagawa, 1972a,b) (Fig. 1K,L). In other biomembranes, inactive vesicular structure can be reconstituted after drastic fractionation (cf. Zahler and Weibel, 1970), but special methods are required to reconstitute biologically active vesicles (Razin, 1972). Before going into specific methods of dissociation and reassociation, it is necessary to describe the essential properties of the components.

1.2. Electron Transport System

1.2.1. Chromoproteins

The electron transport system is composed of many proteins which carry prosthetic groups such as heme, flavin, and nonheme iron. These prosthetic groups as well as coenzyme Q can be oxidized and reduced to transport electrons during the respiration with light-absorption changes specific to each compound (see Section 4.1). Aided by their characteristic light absorption, isolation of cytochromes a, b, c and c_1 has been achieved by many workers since the 1930s (Keilin and Hartree, 1937; Okunuki, 1960; Wainio, 1970). Flavoproteins such as succinate dehydrogenase [succinate:(acceptor) oxidoreductase, E.C. 1.3.99.1] were also highly purified (Davis and Hatefi, 1971).

However, complete reconstitution of the electron transport system from highly purified proteins has never been successful. These purified proteins may be conformationally changed or may have lost essential subunits (prosthetic group) that do not contribute to the characteristic spectra during the isolation procedure. In fact, a new oxidation factor which helps electron transport between cytochromes b and c_1 has recently been discovered (Nishibayashi-Yamashita *et al.*, 1972).

1.2.2. Reconstitution of Electron Transport

Only a limited number of extrinsic proteins (see Section 1.5) such as cyto-chrome c and succinate dehydrogenase (Hanstein *et al.*, 1971a) can be de-tached from the inner membrane and readsorbed on the membrane prep-arations with restoration of electron transport. On the other hand, partial reconstitution of electron transport is also possible by combining crude lipid-containing preparations, i.e., complexes I (Fig. 1F), II, III, and IV (see Table I). The combination of these complexes together with cytochrome

Table I. Fragments of Electron Transport

Fragment	Enzyme name	Composition (μmoles/g protein)	Phospholipid (%)	Specific inhibitor
Complex I	NADH-CoQ reductase	FMN (1.4), nonheme Fe (26)	22	Amytal, rotenone
Complex II	Succinate-CoQ reductase	FAD-His (4.5), heme b (4.6), nonheme Fe (36), heme c (1.5)	20	TTFA[a]
Complex III	CoQH$_2$-cytochrome c reductase	Heme b (8.5), heme c (4.1), nonheme Fe (11)	40	Antimycin
Complex IV	Cytochrome c oxidase	Heme a (8.5), Cu (9.4)	35	CN, CO

[a] Thenoyltrifluoroacetone.

c and coenzyme Q results in the formation of NADH- or succinate-oxidase (Hatefi, 1966). The combination of, for example, complexes II and III (Rieske, 1967) results in the succinate cytochrome c reductase, which is a partial reaction of electron transport (Hatefi *et al.*, 1962). This kind of reconstitution requires only a dilution procedure after mixing complexes in a small volume. The resulting preparation is a membrane and shows a continuous change in activity and density, depending on the ratio in which the complexes have been mixed (Tzagoloff *et al.*, 1967). In this reconstituted

membrane, water-insoluble intrinsic proteins are held together by phospholipids, but the mode of interaction between these proteins may be entirely
different from that between protomers in an oligomer, where protomers
interact with strict stoichiometry and topology.

1.3. Energy Transfer System

1.3.1. ATPase (F_1) and Proteins Conferring Oligomycin Sensitivity on F_1

The energy transfer system is also composed of many proteins, but to date
no prosthetic group has been found in them. For this reason, the isolation
of these components was started only after the late 1950s (Racker, 1970).
These proteins are called "coupling factors" because they are capable of
restoring oxidative phosphorylation in appropriate deficient particles
(Fig. 1D,E) which contain the electron transport system, phospholipids, and
the remaining portion of the energy transfer system (see Section 6). Although
many coupling factors have been described in the literature, only ATPase
(F_1) and oligomycin-sensitivity-conferring protein (OSCP) are pure and
well characterized. Coupling factors 2, 3, 4, 5, and 6 (Racker, 1970), coupling
factors B, C, and D (Refer to Fisher et al., 1971), and factor X or many
others await further identification. Coupling factor 1 is identical to factor A
except for its ATPase activity, and coupling factor 2 and factor B may be
identical (Racker et al., 1970). It appears that the active component of
coupling factors 3, 4, and 5 is, at least in part, OSCP (Senior, 1971). Factor
X is also identified as OSCP (van de Stadt et al., 1972).

Purified mitochondrial ATPase is also called coupling factor 1 (F_1)
because it restores oxidative phosphorylation to F_1-deficient submitochondrial particles (Pullman et al., 1960). F_1 has a molecular weight of about
340,000 (Fig. 1G,H) and is made up of three major subunits (Knowles et al.,
1972) (53,000, 50,800, and 33,000 daltons) (cf. Catterall and Pedersen, 1971)
and a few minor components including an ATPase inhibitor (5570 daltons).
Since the ATPase inhibitor may be removed partially during the preparation
of F_1, the specific activity of ATPase in F_1 can vary from 0 (factor A) to 100
μmoles Pi liberated/mg/min in its highest purity and coupling activity
(Horstman and Racker, 1970).

Coupling factor 4 (F_4) was found to restore phosphorylation (Conover
et al., 1963) and to combine with F_1 (labeled with ^{3}H-acetyl group) on the
alkali-treated deficient particles and render F_1 sensitive to many energy
transfer inhibitors including oligomycin (Kagawa and Racker, 1966a,b).
Further purification of F_4 (Bulos and Racker, 1968) resulted in the single

basic protein of 18,000 daltons called oligomycin-sensitivity-conferring protein (OSCP) (MacLennan and Tzagoloff, 1968).

To confer oligomycin sensitivity on F_1, not only OSCP but also phospholipids and some membrane proteins were shown to be essential (Kagawa and Racker, 1966b). ^{14}C-Rutamycin or ^{14}C-dicyclohexylcarbodiimide (DCCD) was shown to be bound on the membrane portion other than F_1 and OSCP (Kagawa and Racker, 1966b; Cattell et al., 1971). Both agents are also energy transfer inhibitors. The DCCD-binding protein was purified by chloroform–methanol (2:1) extraction and fractionation on Sephadex LH20 as a single protein of about 10,000 mol wt.

1.3.2. Reconstitution of Oligomycin-Sensitive ATPase

F_1, OSCP, and ATPase inhibitor are extrinsic proteins and are soluble in water after their detachment from the membrane. However, when detergents were applied to the membrane, these proteins and some intrinsic membrane proteins were extracted together (Fig. 1I), and the preparation showed oligomycin-sensitive ATPase activity in the presence of phospholipids (Kagawa and Racker, 1966b; Tzagoloff et al., 1968a; Kopaczyk et al., 1968). Although most of the electron transport components were removed from this oligomycin-sensitive ATPase including Swanljung's preparation (1971), more than ten protein subunits were detected in it when the preparation was analyzed by dodecylsulfate gel electrophoresis (Stekhoven, 1972). The absence of electron transport components in oligomycin-sensitive ATPase is established in promitochondria, where no cytochrome, nonheme iron, or coenzyme Q is detected (Groot et al., 1970).

Although reconstitution of oligomycin-sensitive ATPase from completely purified proteins has not been achieved, the following component studies have been accomplished:

1. F_1 plus F_0 (urea-treated submitochondrial particles) (Kagawa and Racker, 1966a).
2. F_1 plus OSCP (chaotropic-alkali treated particles) (Kagawa and Racker, 1966a; MacLennan and Tzagoloff, 1968).
3. F_1 plus phospholipid plus CF_0 (cholate-treated F_0) (Kagawa and Racker, 1966b; Kopaczyk et al., 1968).

The possible role of oligomycin-sensitive ATPase as a H^+-translocating ATPase in oxidative phosphorylation has already been proposed (Mitchell, 1967). In fact, vesicles capable of H^+ accumulation by ATP hydrolysis have actually been reconstituted (Kagawa, 1972a,b). In this case, not only oligo-

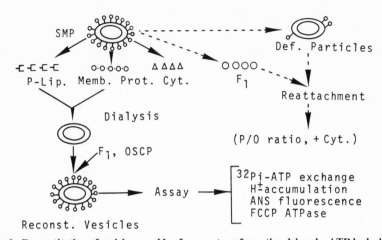

Fig. 2. Reconstitution of vesicles capable of energy transformation driven by ATP hydrolysis.
The old method of reconstitution is indicated by the dashed lines. Coupling factors are
detached from the membrane and reattached on the membrane vesicles to restore net
oxidative phosphorylation (P/O ratio) as described in Section 4.5.3. The new method of
reconstitution is indicated by the solid lines. The vesicular membrane is dissolved and
components are fractionated. Reconstitution is performed by dialysis of membrane
proteins and phospholipids in the presence of detergents. Energy transfer reactions
indicated are assayed after the addition of F_1 and OSCP as described in Section 4.5 and 7.3.
SMP, submitochondrial particles prepared as described in Section 6.1.4; Def. Particles,
deficient particles prepared as described in Section 6.1.1.; P-Lip, phospholipids as described in Section 1.4 and Table XI; Memb. Prot., intrinsic membrane proteins described
in Section 6.3; Cyt., cytochromes a, a_3, b, c, and c_1 described in Section 1.2.1; Reconst.
Vesicles., reconstituted vesicles by the new method as described in Section 6.2.3; F_1,
coupling factor 1 or purified mitochondrial ATPase described in Section 5.1.1; OSCP,
oligomycin-sensitivity-conferring protein described in Section 5.1.2; FCCP, carbonylcyanide p-trifluoromethoxyphenylhydrazone, an uncoupler, ANS, 1-anilinonaphthalene-
8-sulfonate described in Section 7.3.4.

mycin inhibition but also uncoupler stimulation of the ATPase activity is
observed (Kagawa and Racker, 1971) (for details, see Section 7). The
reconstituted vesicles, which lost respiratory activity during the fractionation, cannot be tested by net phosphorylation (or P/O ratio, see Section
4.5.3) like deficient particles, but can be assayed by ^{32}Pi-ATP exchange,
ATP-driven H^+ accumulation (see Section 7.3.3), ANS fluorescence (see
Section 7.3.4), and uncoupler (FCCP) stimulated ATPase (Fig. 2).

These reconstitution studies also confirmed that the 90 Å particles on
the inner membrane are F_1 itself (Fig. 1A–D,F,G) (Kagawa and Racker,
1966c; Racker and Horstman, 1967; Kopaczyk et al., 1968).

1.4. Ion Transport and Membrane Lipids

1.4.1. Phospholipids in the Inner Mitochondrial Membrane

The inner mitochondrial membrane contains lipids as other biomembranes do. Compared with other biomembranes, this membrane is characterized by its low total lipid content (27 % by weight), low cholesterol content (less than 4 % of phospholipids), high cardiolipin content (15 % of total phospholipids), and the presence of coenzyme Q (4 μmoles/g). The most abundant phospholipids are phosphatidylcholine (41 % of the total phospholipids) and phosphatidylethanolamine (33 % of the total phospholipids) (Fleischer et al., 1961). These phospholipids contain unsaturated fatty acyl groups, including linoleic acid and arachidonic acid at the C-2 position. Although the above-mentioned polyunsaturated acyl groups can be completely replaced by monounsaturated ones without impairing oxidative phosphorylation (Kagawa et al., 1969), further saturation of phospholipids causes loss of phosphorylation before the loss of ATPase and respiration (Haslam et al., 1971).

1.4.2. Phospholipids in Reconstitution of Functions

Although restoration of the respiratory activity of lipid-depleted tissues by the addition of phospholipids has been described since the 1920s (Tsuneyoshi, 1927; Kakiuchi, 1927; Fleischer et al., 1961), phospholipid requirements in phosphorylation and H^+ transport have been established only recently (Kagawa et al., 1973).

Reconstitution of mitochondrial function revealed at least two roles of phospholipids in mitochondria: (1) to activate enzymes bound to the membrane structure, perhaps by restoring the original conformation required to interact with their substrates, and (2) to form membrane vesicles tight enough to keep an ion gradient during energy transfer reactions (Fig. 1A–E,K,L, Fig. 2)

The first role is demonstrated by the lipid-depleted fragments. Gentle removal of phospholipids from complexes by 10 % aqueous acetone (Fleischer et al., 1961; Fleischer and Fleischer, 1967) or from oligomycin-sensitive ATPase by 2 % cholate–ammonium sulfate (Kagawa and Racker, 1966b; Berezney et al., 1970) results in a preparation with potential activity. The activity is restored either by pure phosphatidylcholine, phosphatidylethanolamine, or even cardiolipin (Kagawa and Racker, 1966b; Bulos and Racker, 1968; Swanljung, 1971) when fatty acyl groups of these phospholipids are to some extent unsaturated (Kagawa and Racker, 1966b).

1.4.3. Vesicles and Ion Transport

The presence of phospholipids in the abovementioned enzyme preparations does not necessarily mean that vesicular structure is required for their enzyme activities. In fact, some oligomycin-sensitive ATPase or cytochrome oxidase activity can be assayed in the presence of certain amounts of detergents which disperse the preparation into fine fragments. However, vesicular structure is essential for ion transport (*cf.* Kagawa, 1972*a,b*). The role of lipid membranes in ion transport and uncoupler effects has been discussed in detail by Haydon and Hladky (1972).

In the reconstitution of active vesicles, thermotropic and lyotropic mesomorphism (Ladbrook and Chapman, 1969) of phospholipids as well as surface charge of the vesicles are considered. As mentioned earlier, there is a strict requirement for unsaturated acyl groups in phosphorylation. In this reconstitution, phosphatidylcholine and phosphatidylethanolamine are both required, and cardiolipin is also needed for optimal activity (Kagawa *et al.*, 1973). As to the lyotropic mesomorphism during the reconstitution, phospholipids should be brought into isotropic solution from the lamellar phase (Fig. 1J) before intrinsic proteins are added (Kagawa, 1972*b*). These requirements are more strict than those needed to reconstitute mere enzyme activity.

The reconstituted vesicles can transform energy of oxidoreduction (Hinkle *et al.*, 1972) or ATP hydrolysis (Kagawa *et al.*, 1973) into ion transport; these processes are sensitive to uncouplers. The uncouplers are considered to be lipid-soluble weak acids. Both protonated and dissociated acid molecules can go through the lipid membrane of vesicles, according to the electrochemical potential of the protons and thus are able to carry H^+ between the inside and the outside of the vesicle and to dissipate the energy stored in the vesicular structure. As predicted by Mitchell (1966) and Skulachev (1970) for the mitochondria and submitochondrial particles, the reconstituted vesicles also react with lipid-soluble ions such as K^+-valinomycin complex or tetraphenyl boron which move according to the membrane potential (Kagawa and Racker, 1971).

1.5. Topology in the Membrane: Extrinsic and Intrinsic Proteins

The structure of biological membranes is not uniform. It became clear that some proteins are extrinsic or attached on the surface of the membrane and others are intrinsic or inlaid in the membrane. Generally, extrinsic proteins are detached by sonic oscillation or treatment with salt, alkali, chaotropic anions, or chelating agents, while intrinsic proteins are solubilized by detergents or other agents which dissolve phospholipids. Extrinsic proteins are water soluble and easily isolated by the usual enzyme purification procedures such as ammonium sulfate fractionation or ion-exchange chromatography (see Sections 2 and 5). There is no established way of fractionating intrinsic proteins in their native and active forms, owing to their hydrophobic properties, but several methods are described in Section 6.

The anisotropic arrangement of inner membrane components has been demonstrated (Chance *et al.*, 1970; van Dam and Meyer, 1971; Kagawa, 1972*a*). As shown in Table II, chemical labeling with ^{35}S-diazobenzene sulfonate from outside (mitochondria) or inside (submitochondrial particles) gave entirely different incorporation patterns of ^{35}S (Schneider *et al.*, 1972).

Table II. Radioactivity of Components of Mitochondrial Inner Membrane after Labeling with ^{35}S-Diazobenzene Sulfonate^a

Components	^{35}S incorporated (cpm/mg protein)	
	Submitochondrial particles	Mitochondria
F_1	1530.0	84.4
Cytochrome *c*	26.0	836.0
Cytochrome oxidase	5.9	24.5
Cytochrome *b*	12.7	40.7

^{a} The condition of the labeling was essentially the same as reported by Schneider *et al.* (1972). Purified F_1, cytochrome *c*, and cytochrome oxidase were labeled equally, (about 1000 cpm/mg) in the comparable experiments.

This kind of anisotropy is essential in the reconstitution of phosphorylating vesicles. For example, in the reconstitution of phosphorylation coupled with cytochrome oxidase, cytochrome *c* must be localized inside the vesicles (Racker and Kandrach, 1971).

1.6. The Scope of This Chapter

Thousands of original papers have been published on this subject, especially on the electron transport system (Wainio, 1970) and on general membrane systems (Razin, 1972). However, only a few reports have been published on the reconstitution of an energy transfer system from solubilized components.

For this reason, experimental details are given only for the reconstitution of energy transfer of mammalian mitochondria (Kagawa, 1972a).

Composition and physical constants of solutions or reagents discussed in this chapter can be found in a handbook (Sober, 1970). General preparative methods and apparatus (Jakoby, 1971) and those specific for mitochondrial fractions (Estabrook and Pullman, 1967; Okunuki, 1960) have already been described. Methods of dissociation (see Section 2), fractionation (see Section 3), and analysis (see Section 4) of the inner membrane components are limited to the case where special comments are needed to handle membranes, since general methods developed for soluble proteins are often difficult to apply to the membrane components. With regard to the extrinsic proteins, preparative methods are described only where they yield highly pure proteins with known molecular weight.

2. METHODS OF DISSOCIATION OF THE INNER MEMBRANE

2.1. Mechanical Disintegration

2.1.1. Sonic Oscillation of the Membrane

Mechanical disintegrations such as sonic oscillation, homogenization, freezing-thawing, osmotic shock, Nossal shaking, and bursting with a French press are applied to isolate mitochondria from the cells, to separate inner membrane from outer ones, or to disrupt mitochondria into submitochondrial particles (Fig. 1B,C). However, stronger conditions are required to dissociate membrane components by mechanical methods. For example, to disrupt mitochondria into submitochondrial particles, 5 min sonication in a Raytheon sonicator at 10 kc/sec at maximum output at 0°C is enough (Kagawa and Racker, 1966b), while to detach ATPase (Fig. 1G) from them 20 min sonication with a Branson sonifier (temperature is raised to 55°C) is required (Horstman and Racker, 1970). Although sonication is a convenient and effective method, high-speed shaking with glass beads in a Nossal shaker is sometimes used (Penefsky et al., 1960).

2.1.2. Cavitation

Theories of sonic treatment have been reviewed by several authors (Peacocke and Pritchard, 1968; Hughes and Nyborg, 1962). Although the lower frequency limit of ultrasonic waves is defined as 16 kc/sec, mitochondrial membranes are, practically, disrupted either by 10 kc/sec (Raytheon) or by 90 kc/sec (Sonblaster model G201) at an input energy of about 100 W. The intensity of plane sound waves (I) decreases with the distance (d) from the source according to

$$I = I_0 e^{-2\alpha d}$$

where I_0 is the sound intensity at $d = 0$ and α is the absorption coefficient. The volume of the sample is limited for this reason, and the viscosity of the mitochondrial suspension is lowered by diluting the protein concentration to less than 30 mg/ml to decrease α value.

The most important factor in this treatment is "cavitation." The cavities are formed in a liquid when the liquid is exposed to ultrasound of sufficient intensity to cause large local pressure during the negative-pressure phase. The onset of cavity formation is accompanied by an audible hissing noise and appearance of fine bubbles. In the collapse phase, instantaneous temperature of the order of $10^4\,°K$, pressure of the order of 10^6 atm, and free radicals are produced; these are supposed to disrupt biomembranes. The sizes of the bubbles are varied, but the stable resonance radius of air bubbles is given by the equation

$$a = \frac{3.0}{f}$$

where f is the frequency of the ultrasound (kc/sec) and a is the radius (mm). For 20 kc/sec, the diameter is 150 μ. The minimum pressure amplitude which will produce the cavities is called the "cavitation threshold." If the liquid contains dissolved gases, this threshold is considerably lowered. So sonication in N_2 is more effective than that in vacuum to avoid oxidation. To prevent the damaging effect of free radicals, SH compounds such as dithiothreitol are added as scavengers.

2.1.3. Apparent Solubilization by Sonication

Extensive sonication, especially at high pH, seems to "solubilize" even intrinsic proteins. However, these solubilized preparations are composed of small fragments of different sizes but of the same cytochrome content (Tzagoloff et al., 1968b). Soluble ATP synthetase composed of two pure proteins was reported to be purified from the sonic extract of mitochondria without detergent (Fisher et al., 1971); however, the preparation is composed of 14 proteins (Stekhoven, 1972) and may contain fine vesicles (see Section 3.1).

Since intrinsic proteins and phospholipids are physically insoluble in the water phase, destruction of membrane structure by itself does not cause solubilization. On the other hand, extrinsic proteins which are rich in hydrophilic groups and arranged on the surface can stay in proper solution if they are detached by sonication.

2.2. Chaotropic Agents

2.2.1. Properties of Chaotropic Anions

Chaotropic anions are the group of ions which have a large diameter, single charge, and large positive entropy of aqueous ions. Examples are SCN^-, I^-, and haloacetates. The destruction of water structure by these anions weakens hydrophobic bonds by raising the solubility of hydrophobic amino acids and thus causes dissociation of some proteins from the mitochondrial membrane (Hanstein et al., 1971b). Guanidine-HCl and urea have similar effects and at 3 M increase the solubility of acetyltetraglycine ethyl ester (peptide analogue) 3.5- and 1.8-fold, respectively. The solubilizing effect and protein-dissociating power of urea are greater at 0 °C than at 25 °C. The order of effectiveness of these agents in extracting proteins from membrane is $CCl_3COO^- > SCN^- >$ guanidine $> ClO^-_4 > Br^- > NO_3^-$ $>$ urea. The dissociation of membrane-bound enzymes such as ATPase (Kagawa and Racker, 1966a,b), NADH dehydrogenase (Hanstein et al., 1971b), succinate dehydrogenase (Davis and Hatefi, 1971), and cytochrome c_1 (Rieske et al., 1967) was accomplished by these agents. The advantage of these reagents over detergents is their easy removal, while a disadvantage is their denaturing effects on the extracted proteins. In fact, these agents are used to remove some proteins from membrane, but the resulting extract is, in many cases, not useful for the preparation of active enzymes.

Practically no phospholipids or intrinsic proteins are extracted, because of the weak solubilizing effect. Chaotropic extraction was successful in the preparation of succinate dehydrogenase (Davis and Hatefi, 1971) and oxidation factor (Nishibayashi-Yamashita et al., 1972).

The succinate dehydrogenase detached by chaotropic agent could be readsorbed on complex II if an antichaotropic anion was added (Davis and Hatefi, 1972). Antichaotropic anions are large polyvalent or small monovalent anions such as SO_4^{2-}, HPO_4^{3-}, and F^- and are water structure formers. The role of antichaotropic anions in reconstitution is discussed in Section 7.

2.3. Detergents

2.3.1. Detergents and Energy Transfer System

Many detergents are used to extract membrane proteins (Hatefi, 1966; Uesugi *et al.*, 1971; Okunuki, 1966), but only a limited number have proved to be suitable for resolution of energy transfer systems. As discussed in detail by Kagawa (1972a), energy transfer reactions such as H^+ transport and ATP synthesis are much more sensitive to the detergent than electron transport reactions. About thirty- to a thousandfold more detergent is required to solubilize membrane protein than to uncouple oxidative phosphorylation. Reconstitution of electron transport is accomplished by mere dilution (Hatefi *et al.*, 1962), while reconstitution of energy transfer requires special methods (Kagawa, 1972a). The reason why a trace amount of detergents can uncouple phosphorylation is not clear yet, but detergents can interact with phospholipid bilayers and mobilize phospholipid molecules, and thus tight retention of H^+ in the vesicular structure becomes impossible. Leakage of H^+ is known to uncouple phosphorylation (Mitchell, 1966). The tightness of the membrane can be measured by NMR (see Section 7) and other physical methods (Ladbrook and Chapman, 1969).

Mild detergents such as digitonin have been used to disperse membranes, but there is no separation of components, i.e., true solubilization, by digitonin (Hoppel and Cooper, 1968).

2.3.2. Removable Detergents and Their Physical Properties (HLB, cmc, and Aggregation Number)

As discussed above, the removal of the detergent after reconstitution is essential. For this purpose, the following physical properties of the detergent may be considered: (1) It should have a high hydrophilic–lipophilic balance (HLB > 15) so as to be soluble in water. (2) Its critical micelle concentration (cmc > tens of mM) should be high in order to prevent its tight binding to the membrane. (3) Its aggregation number and micelle weight (aggregation number × molecular weight of the detergent < a few thousand daltons) should be small.

HLB is the ratio of hydrophilicity and lipophilicity within a molecule, as defined by the equation

$$HLB = 0.36 \ln \left(\frac{C_o}{C_w} \right) + 7$$

where C_o/C_w is the distribution ratio of the detergent molecule between hydrocarbon oil and water. C denotes the concentration, and the subscripts o and w refer to oil and water (Davis, 1957). In Table III, HLB is related to the solubility and applications of the detergents. The critical micelle concentration (cmc) is the concentration of the detergent at which micelle formation begins in solution. Around the cmc, the physical properties of the solution are dramatically changed and the solubilizing effect is increased. The cmc is determined not only by the properties of the detergents but also by the condition of the solvent. The addition of salt decreases cmc;

Table III. Hydrophilic–Lipophilic Balance (HLB) of Detergents and Their Application

HLB	Compounds (commercial name)	Applications or effects	Solubility of detergents
0	Hexane, paraffin		Insol.
1	Oleic acid		Hd. disp.[a]
2	Sorbitan trioleate (Soan 85)	Defoaming	Hd. disp.
3	Sorbitol hexastearate adduct[b] (G1050)	Defoaming	Sl. disp.[c]
4	Monostearin	Emul. (w/o)[d]	Sl. disp.
5	Sorbitan monostearate (Emasol 310)	Emul. (w/o)	Sl. disp.
6	Diethylene glycol monolaurate (G2141)	Emul. (w/o)	Sl. disp.
7	Sorbitan monopalmitate (Span 40)	Wetting	Sl. disp.
8	Tetraethylene glycol monooleate (G2140)	Wetting	Milky disp.[e]
9	Sorbitan monolaurate (Span 20)	Wetting	Milky disp.
10	Hexaoxyethylene oleyl ether (Emulgen 408)	Emul. (o/w)[f]	Milky disp.
11	Sorbitan trioleate adduct[b] (Tween 85)	Emul. (o/w)	Trans. disp.[g]
12	t-Octyl phenol adduct[b] (7.5) (Triton X-114)	Emul. (o/w)	Trans. disp.
13	Sorbitol lanolin adduct (G1431)	Emul. clean[h]	Trans. disp.
14	t-Octyl phenol adduct[b] (9.5)	Emul. clean	Clear soln.
15	Poly(20)oxyethylene sorbitan monooleate (Tween 80)	Emul. solb.[i]	Clear soln.
16	Poly(20)oxyethylene palmityl ether (Briji 58)	Emul. solb.	Clear soln.
17	Poly(23)oxyethylene lauryl ether (Briji 35)	Emul. solb.	Clear soln.
18	Poly(40)oxyethylene stearate (Myrji 52)	Emul. solb.	Clear soln.
20	Sodium oleate	Emul. solb.	Clear soln.
40	Sodium dodecylsulfate	Emul. solb.	Clear soln.

[a] Hd. disp., hardly dispersible.
[b] Adduct (n), adduct of polyoxyethylene (number of polyoxyethylene groups in the detergent molecules).
[c] Sl. disp., slightly dispersible.
[d] Emul. (w/o), emulsification as water-in-oil type emulsion or suspension.
[e] Milky disp., milky dispersion.
[f] Emul. (o/w), emulsification as oil-in-water type emulsion or suspension.
[g] Trans. disp., transparent dispersion.
[h] Emul. clean., emulsification (o/w) and cleansing.
[i] Emul. solb., emulsification (o/w) and solubilization of hydrophobic substance in water phase.

moreover, it disrupts electrostatic bonding in the biomembrane. For this reason, 0.1–0.5 M salt is added during the detergent extraction of biomembranes. In extraction of membrane components, a detergent concentration above cmc is used, while in reconstitution, a dialysis mixture containing a detergent concentration lower than cmc is better (Kagawa et al., 1972).

2.3.3. Solubilization of Phospholipids Before the Reconstitution

The other important factor of detergents in extraction and reconstitution of oxidative phosphorylation is their ability to dissolve phospholipids. Since phospholipids are amphiphiles and their cmc is extremely low (on the order of 10^{-5} to 10^{-10} M), it is difficult to pull single molecules out of phospholipid bilayers or other polymorphic states (Gulik-Krzywicki et al., 1967; Kamat and Chapman, 1970; Ladbrook and Chapman, 1969). A detailed phase diagram of cholate–phosphatidylcholine–water (Small et al., 1966; Hofmann and Small, 1967) shows that at least 1 mole of cholate per 2 moles of phosphatidylcholine is required to bring the phosphatidylcholine molecules from the lamellar phase into isotropic solution (see Section 7.1).

2.3.4. Complete Solubilization of Biomembranes

Complete solubilization of biomembranes into subunits is possible by the use of sodium dodecylsulfate (1.4 g/g protein) in the condition described by Tanford (1972), where the protein subunit becomes a rod-shaped complex (see Section 4.2). Although it is not yet successful in the reconstitution of biomembranes, subunits of soluble enzymes thus obtained can be reassociated into an active oligomer if dodecylsulfate is removed by Dowex-1 in the presence of urea (Weber and Kuter, 1971).

2.3.5. Effective Use of Bile Acids

Bile acids are flat molecules with OH and COO$^-$ on one face and CH$_3$ on the other face, and can thus cover large hydrophobic areas without impairing protein conformation. The detergent properties of bile salts have been described in detail (Hofmann and Small, 1967). Cholate has been successfully used in the preparation of cytochromes (Okunuki, 1960), complexes of electron transport (Rieske, 1967; Hatefi, 1966), oligomycin-sensitive ATPase (Kagawa and Racker, 1966b; Tzagoloff et al., 1968a), and exchange particles (Arion and Racker, 1970) and in the reconstitution of oxidative phosphorylation (Kagawa, 1972a; Kagawa et al., 1973; Racker and Kandrach, 1971) because of its high HLB (18), high cmc (45 mM), and low aggregation number (2). Deoxycholate is about fourfold stronger in its solubilizing effect (Wainio, 1970) because of its low cmc (10 mM). However, ammonium sulfate fractionation in the presence of deoxycholate is difficult (see Section 3.3).

Nonionic detergents such as Triton X-100 are also used to extract membrane components; the extract can be fractionated on an ion-exchange column. However, the removal of nonionic detergents after extraction is difficult because of their low cmc and high aggregation number.

2.4. Organic Solvents

2.4.1. Dielectric Constant and Extracting Effect of Solvents

Since organic solvents are easily removed after the treatment and show more powerful dissolving capacity than detergents, these are the agents of choice if they are applicable. Polar organic solvents with intermediate dielectric constant (between 4 and 39) can dissolve lipids and some hydrophobic peptides. As shown by Kakiuchi (1927) and Tsuneyoshi (1927), reconstitution of electron transport can be accomplished with these solvents. However, they denature most of the membrane proteins of energy transfer. If the solvent exhibits a dielectric constant over 40, as do dimethylsulfoxide and ethylene glycol, practically no lipids are extracted, and phosphorylating activity is protected by these at 10 % (v/v). If the solvent exhibits a dielectric constant of less than 1.9, such as hexane does, the effect on phosphorylation is also small.

2.4.2. Conditions of Extraction

As studied in detail by Lenaz et al. (1972), there are rules to predict the effect of organic solvents. The efficiency of phospholipid extraction is not affected by temperature (0 °C, 20 °C, 32 °C) and is not different in mitochondria and submitochondrial particles. The addition of chaotropic anion, acid, or alkali helps the extraction with solvents. There is a linear relationship between the logarithm of the reciprocal concentration of the alcohol required to extract 50 % of the phospholipids and chain length of the alcohol [log (1/concentration) $\propto$ carbon chain length]. These concentrations are 18 % methanol, 11 % ethanol, 2.7 % n-propanol, and 1.4 % n-butanol. Salt solutions containing t-amyl alcohol or n-butanol have been used to extract electron transport components. Extraction with organic solvents such as 10 % aqueous acetone removes most of the phospholipids from mitochondria or the complexes of electron transport, and the electron transport activity of these residues is restored by the addition of a dispersed phospholipid suspension (Fleischer and Fleischer, 1967).

The isolation of DCCD-binding protein (see Section 6.3) from mitochondria (Cattell *et al.*, 1971) or from oligomycin-sensitive ATPase (Stekhoven *et al.*, 1972) was successful by extraction with a chloroform–methanol (2:1) mixture followed by fractionation on a Sephadex LH20 column. The damaging effect of this mixture on the energy transfer system may be avoided if the extraction is performed at $-75\,°C$ as in the case of Na-K ATPase (Noguchi and Freed, 1971).

2.5. Alkali, Salt, Chelating Agent, and Other Agents

Extraction of mitochondria with alkali-salt releases the crude protein mixture called "F_4" (Conover *et al.*, 1963). F_4 binds F_1 and confers oligomycin sensitivity on F_1, which is bound to the alkali-extracted membrane fragments (see Section 6.3) (Kagawa and Racker, 1966a; Bulos and Racker, 1968). The active component of F_4 was purified and renamed "OSCP" by MacLennan and Tzagoloff (1968) (see Section 5.1).

Salt breaks electrostatic bonds between cytochrome *c* and mitochondrial membrane. The surface of cytochrome *c* molecule is covered with both acidic and basic amino acid residues (Dickerson *et al.*, 1971).

Chelating agent is added to remove divalent cations which presumably connect some proteins to the membrane. F_1 is detached by mechanical methods in the presence of 2 mM EDTA without salt (Pullman *et al.*, 1960).

Phospholipases have been used to remove phospholipids from submitochondrial particles, and the resulting fragments lost the ability to confer oligomycin sensitivity on F_1 (Kagawa and Racker, 1966a; Fleischer and Fleischer, 1967; Berezney *et al.*, 1970) or oxidative phosphorylation (Burstein *et al.*, 1971). The activity lost by phospholipase A treatment can be restored after stopping the enzyme reaction by EDTA to remove Ca^{2+}, if the resulting fatty acids and lysophospholipid are removed by albumin and if phospholipids are supplied (Kagawa and Racker, 1966a). Phospholipase A treatment also releases NADH dehydrogenase from submitochondrial particles (Singer, 1966).

F_1, OSCP, and F_6 are removed by 1% silicotungstate (4°C, 6 min, pH 7.5) from submitochondrial particles (Racker *et al.*, 1969). The mechanism of this agent is unknown, but it is interesting that silicotungstate can precipitate nonionic detergents such as Triton by forming a polyionic complex.

3. METHODS OF FRACTIONATION OF MEMBRANE PROTEINS

3.1. Centrifugation

3.1.1. Centrifugation of Membrane Fragments

Centrifugation is widely used to isolate mitochondria (Hogeboom *et al.*, 1948; Blair, 1967) or their components (Hatefi, 1966; Okunuki, 1960). The centrifugation of membrane preparations is quite different from that of soluble proteins. These membrane preparations are:

1. Heterogeneous in their size and shapes, as shown by electron microscopy (Fig. 1B–F).
2. Larger than common soluble proteins (10,000–1,000,000 daltons), short-time ultracentrifugation being often used for these preparations.
3. Lower in density than pure proteins owing to their phospholipid contents.
4. Labile and easily affected by the presence of detergents, salts, and other agents.

The supernatant obtained by centrifugation in a Spinco rotor 50Ti at 50,000 rpm for 1 hr is often considered to be soluble, or free of membrane. However, to establish solubility there are many physical properties which are essential, as discussed in the following section.

3.1.2. Performance Index

Usually, the condition of centrifugation is expressed as the relative centrifugal force (G in g) and the time of centrifugation:

$$G = 1.10 \times 10^{-5} \times R \times \text{rpm}^2 \ (g), \ g = 980 \ \text{dyn}$$

where R is the average distance of the contents of the centrifuge tubes from the axis of rotation. Mitochondria are precipitated at 17,000 rpm in a Spinco 30 rotor ($R_{max} = 10.0$ cm, $R_{min} = 5.0$ cm, $79,000 \times g$) for 10 min (Blair, 1967), while submitochondrial particles are sedimented at 40,000 rpm in a Spinco 40 rotor ($R_{max} = 8.1$ cm, $R_{min} = 3.8$ cm, $105,000 \times g$) for 45 min (Kagawa, 1967*a*). Here, R_{max} and R_{min} are the distances in centimeters from the axis of rotation to the bottom of the tube and to the meniscus (surface), respectively. However, for more subtle fractionation, the volume of the tube content and other factors should be considered. Precipitation of a particle in a centrifuge tube from the top (R_{min}) to the bottom (R_{max}) is determined by the product of two terms, one determined by the rotating rotor (performance index) and the other by the particle in the solvent (sedimentation velocity). The performance index (PI) is defined as follows:

$$PI = \frac{rpm^2}{\ln R_{max} - \ln R_{min}}$$

The 50Ti rotor ($R_{max} = 8.1$ cm, $R_{min} = 3.8$ cm) shows PI $= 33 \times 10^8$, if the tubes are filled to the top and spun at 50,000 rpm.

3.1.3. Precipitation Time

The sedimentation velocity (s) of the particles is determined as follows if the particle is spherical:

$$s = \frac{(\sigma - \rho)d^2}{18\eta}$$

where η is the viscosity of the solvent, d is the diameter of the particles, and σ and ρ are the densities of the particles and the solvent, respectively. Then the time (T_p in min) required to precipitate all the particles is obtained as follows:

$$T_p = \frac{1}{s} \frac{1}{PI} \frac{60}{4\pi^2} = 27.4 \times \frac{1}{PI} \frac{\eta}{d^2(\sigma - \rho)}$$

In the usual centrifugation where water (20°C) is used as the solvent ($\eta = 0.01$ poise, $\rho = 1.0$ g/ml) and the particles are spherical ($\sigma = 1.2$ g/ml), T_p is as follows:

$$T_p = 1.37 \times \frac{1}{PI} \times \frac{1}{d^2}$$

3.1.4. Application to the Membrane

To apply the above equation to specific examples: where the viscosity is increased by 5%, T_p should be increased by 5%; where the density difference ($\sigma - \rho$) is decreased to 50% (σ of mitochondria is 1.1), T_p should be increased twice; if the long axis of the particle is ten times longer than the short axis, T_p is increased to 50%.

In conclusion, centrifugation of a membrane preparation ($\sigma = 1.1$) in a sucrose solution at 0°C in the 50Ti rotor at 50,000 rpm does not remove particles with diameters smaller than 300 Å, which is large enough to accommodate vesicles, composed of a lipid bilayer (80 Å thick) (cf. Section 2.1.3).

On the other hand, even small vesicles such as submitochondrial particles are precipitated at 10,000 $\times$ g for 10 min in the presence of Mg^{2+}, salts, or dilute acids (pH 6), owing to aggregation.

3.1.5. Fractionation by Buoyant Density

Since disruption of a biomembrane results in fragments of different sizes and shapes, fractionation by buoyant density in sucrose solution is sometimes more effective than that by sedimentation velocity. For example, isopyknic centrifugation separates inner membrane (σ = 1.21) from outer membrane (σ = 1.13) of mitochondria. Separation of lipid-containing vesicles and lipid-depleted fragments can be performed by this method.

3.1.6. Short-Time Ultracentrifugation

Short-time ultracentrifugation is often used in fragment preparations such as complex III (30,000 rpm, 15 min; Rieske, 1967) to prevent denaturation during the centrifugation. In order to obtain reproducibility, not only the volume, which affects the performance index, but also the acceleration time should be adjusted to the original conditions. A large rotor or a titanium rotor is much slower in acceleration than the usual one. Recently, the Spinco L3-type centrifuge has been widely used because of its high acceleration rate (10,000 rpm increment/min for the 50Ti rotor).

3.2. Gel Filtration

3.2.1. Gels

A mixture of proteins can easily be separated into individual proteins according to their molecular weight if the mixture is applied to a column of gel. Gels used in this chromatography consist of four types: (1) agarose (Bio-Gel A and Sepharose), (2) dextran (Sephadex), (3) glass (Bio-Glass), (4) polyacrylamide (Bio-Gel P).

The theoretical correlation of molecular Stokes radius with elution position from the gel filtration column is described by Ackers (1967). Practical uses of gels have been described in detail by Reiland (1971).

3.2.2. Elution of Column

The gel equilibrated with desired buffer is packed in a column, and the applied sample is eluted with the buffer under hydrostatic pressure of 30–100 cm H_2O. When a soft or less cross-linked gel such as Sephadex G200 or Agarose A150m is used, it is desirable to reduce the pressure by using a

reverse-flow column or large siliconized glass beads in the column as described
by Sachs and Painter (1972).

3.2.3. K_{av} and K_d

The rate of migration of a protein molecule through the column bed can be
characterized by its distribution coefficient (K_d or K_{av}):

$$K_d = \frac{V_e - V_0}{V_i}$$

where Ve is the elution volume of the sample, V_i is the internal volume, and
V_0 is the void volume. Since V_i is difficult to measure, K_{av} is often used,
which is based on V_t (total column volume):

$$K_{av} = \frac{V_e - V_0}{V_t - V_0} \qquad K_{av} = \frac{K_d}{1 + V_g/V_i}$$

where Vg is the volume of gel material in the column.

3.2.4. Selection of Gels for Membrane Components

An appropriate gel is chosen according to the fractionation range. In
Table IV, properties of gels used for large macromolecules are given. To
obtain V_0 in these gels, blue dextran 2000 (0.5%) may be too small, and
larger particles such as certain viruses (detected by absorption at 260 μ)
are used.

To determine molecular weight, K_{av}s are obtained on both the samples
and several enzymes of known molecular weight, which serve as standards.
Sepharose shows a wide fractionation range of molecular weights, while
Bio-Gel P shows a narrow range of fractionation and thus exact molecular
weight.

3.2.5. Particle Weight of Membrane

Particles may be retained on the column, not because they are really small
but because they are adsorbed on the gel by ionic or other interactions.
Submitochondrial particles are easily aggregated in the presence of salt and
retained. The addition of a detergent in the column helps fractionation. The
criteria of membrane fractionation in this method include (1) chemical analysis, e.g., separation of phospholipid phosphorus from proteins, (2) appearance
of different protein subunits at different K_{av}s and analysis of these subunits
by dodecylsulfate gel electrophoresis of the fractions (Weber and Osborn,

Table IV. Properties of Gels Used for Fractionation of Membrane Fragments

Gels	Water regain (ml/g dry gel)	Fractionation range for globular protein (daltons)	Bed volume (mg/g dry gel)
Sephadex G50	5	1,500–30,000	10
G100	10	4,000–150,000	17
G200	20	5,000–800,000	35
Bio-Gel P100	8	40,000–100,000	22
P200	14	80,000–300,000	47
P300	22	100,000–400,000	70
Sepharose 6B	—[a]	10,000–4,000,000	(6%)[b]
4B	—[a]	20,000–10,000,000	(4%)[b]
2B	—[a]	30,000–40,000,000	(2%)[b]
Bio-Gel A15m	—[a]	40,000–15,000,000	(4%)[b]
A50m	—[a]	100,000–50,000,000	(2%)[b]
A150m	—[a]	1,000,000–150,000,000	(1%)[b]

[a] Gels should be kept wet.
[b] Percent agarose in gels.

1969) (see Section 4.2), and (3) identical K_{av} in the rechromatography of one fraction to the K_{av} of the first chromatography.

Even if these criteria are satisfied, the molecular weight determination may not be correct, since the molecular shape of dissolved proteins may not be identical to that of the standard globular proteins. In fact, there are two- to fourfold differences in the fractionation range of Sephadex G200 between compact globular proteins and randomly extended dextran molecules.

3.3. Salting Out in the Presence of Detergents

3.3.1. Salting Out of Soluble Proteins

The antichaotropic salt (see Section 2.2) can be used to salt out proteins if it is highly soluble, as ammonium sulfate is (saturated at 761.4 g/liter). As described by Cohn (1925), there is a simple relation between the solubility of a protein (S) and the concentration of the salt (u):

$$\log S = \beta - K_s \cdot u$$

where β and K_s are constants for a given protein in a given solution.

3.3.2. Effect of Salts on the Detergent

Insoluble intrinsic proteins can be fractionated in the presence of detergents and salt. In the presence of 1–2% sodium cholate, most of the mitochondrial membrane proteins can be precipitated between 30 and 50% saturation, and the remaining supernatant contains phospholipids and cytochrome c (Okunuki, 1960; Kagawa, 1967a). In 2% cholate and 0.1 M salt, cytochrome oxidase is in larger and heavier fragments and so is easily removed by high-speed centrifugation (see Section 6.2). Sodium deoxycholate (0.1–1.0%) is more hydrophobic than sodium cholate and is precipitated by ammonium sulfate at 10% saturation. In this case, both membrane proteins and phospholipids are floated on the surface of the centrifuge tube. The fractionation of membrane proteins with deoxycholate is possible in the presence of 1 M KCl (see Section 7.2) or with ammonium acetate (Rieske, 1967). In both cases, the fragments still contain phospholipids.

3.4. Ion-Exchange Chromatography

3.4.1. Cellulose Ion Exchangers and Extrinsic Proteins

Cationic ion exchangers such as carboxymethyl (CM) cellulose and anionic ion exchangers such as diethylaminoethyl (DEAE) cellulose have been used successfully in the purification of soluble proteins as well as detached extrinsic proteins. Acidic proteins such as F_1 are adsorbed on DEAE-cellulose and basic proteins such as OSCP are adsorbed on CM-cellulose. The adsorbed proteins are usually eluted with 50–200 mM salt solution.

The contaminating submitochondrial particles in the soluble proteins are negatively charged by their acidic phospholipids and mostly adsorbed at the top of DEAE-cellulose column. However, small amounts of particles which are covered with F_1 are eluted from the column at a salt concentration of 0.15 M, when most of F_1 is already eluted (cf. Fisher et al., 1971).

There is a newly developed dipolar ion exchanger (Porath and Fornstedt, 1970) which is gentle for proteins.

3.4.2. Ion-Exchange Chromatography of Intrinsic Proteins

Nonionic detergents such as Triton X-100 (1%) can be used to fractionate intrinsic proteins with DEAE-cellulose column chromatography. However, ionic detergents such as bile acids, which are used for the same purpose, give poor results, although the detergent ions can serve as the ions in the elution.

Although hydrophilic resins such as cellulose are useful, lipophilic resins can be prepared by cross-linking fatty acyl groups onto amino groups of Bio-Rex 70 (polyacrylic acid resin), and this column (resin linked with oleyl group) is useful in the fractionation of the insoluble cytochromes of yeast mitochondria dissolved in deoxycholate (Weiss and Bücher, 1970).

3.5. Dialysis

3.5.1. Controlled Dialysis for Fractionation of Membrane Proteins

Dialysis is the diffusion of solvents across a porous membrane such as cellophane tubing under a concentration gradient between the inside and outside compartments. The theories and the detailed methods have been described by Craig (1960) and Jacoby (1971).

In the case of fractionation experiments, controlled removal of salts and detergents is often required. For example, to prepare complex I or oligomycin-sensitive ATPase (Tzagoloff et al., 1968a; Swanljung, 1971) 2–3 hr of dialysis is essential to precipitate the cytochrome b-c_1 rich fraction, while to remove sodium dodecylsulfate 3 days of dialysis is required (Razin, 1972).

In the author's experience, this procedure often determines the reproducibility of the preparation, since too much removal of detergents causes precipitation of all membrane proteins.

3.5.2. The Velocity of Dialysis

The velocity of removal of a solute is expressed as $t_{1/2}$ (50% escape time), which is determined by the properties of the cellophane tubing such as thickness, porosity, and size; by the properties of the solute such as molecular weight and aggregation numbers; and by the properties of the dialyzing environment including pH, temperature, and stirring velocity.

Two cellophane tubings widely used are obtained from the Visking Co. (Chicago); the large one has a wall thickness of 0.0008 inch and an inflated diameter of $5/8$ inch, while the small one is 0.002 inch thick and $1/4$ inch in diameter. When 1 ml of sodium ^{14}C-cholate is dialyzed at pH 8.0 and 4°C against 100 vol buffer at the stirring velocity of 60 rpm, the initial $t_{1/2}$ of the large tube is 75 min and that of the small tube is 150 min (Kagawa et al., 1973). The $t_{1/2}$ of ammonium sulfate is about one-third the $t_{1/2}$ of cholate (Kagawa and Racker, 1971). If the temperature is raised 10°C, $t_{1/2}$ is decreased by 30%. There is practically no decrease in $t_{1/2}$ by increasing stirrer velocity higher than 60 rpm or by changing the 100 vol of dialyzing solution before the $t_{1/2}$.

The $t_{1/2}$ of small proteins (less than 10,000 daltons) and peptides is discussed by Craig (1960), and in fact there is some loss in small proteins in the presence of detergents. The $t_{1/2}$ of detergents and amphiphiles is not determined by their molecular weight but by their aggregation number. The $t_{1/2}$ of deoxycholate is a little larger than that of cholate in alkaline condition, while that of Triton, digitonin, and many other detergents is extremely large.

Because of their extremely low cmc (10^{-5} to 10^{-10} M), phosphatidyl-choline and phosphatidylethanolamine (700–800 mol wt) are not dialyzed, even when dialyzable detergents are added inside the dialysis bag, and they form multilayered lammellar phases (Fig. 1J) if reacting protein is absent (Kagawa and Racker, 1971). The effect of the detergent solution is decreased if salts are dialyzed away more rapidly (see Section 2.4). In this case the appropriate concentration of the salt can be added to the dialyzing solution.

3.6. Gel Adsorption

3.6.1. Gel Adsorption of Membrane Proteins

As with many soluble enzymes, gel adsorption has been widely used to purify mitochondrial particles. Cytochrome c was purified with calcium phosphate gel by Keilin and Hartree (1937).

To fractionate intrinsic proteins, the addition of detergents such as 1 % cholate during this procedure is necessary. The crude protein mixture is added to Cγ gel or calcium phosphate gel (1 mg dry gel/1 mg protein) in the presence of dilute buffer at 0 °C for 10 min with stirring and eluted with 0.1–0.8 M phosphate buffer at pH 7.0–9.0. The remaining proteins with the gel are easily removed by centrifugation at 10,000 × g for 5 min.

4. ANALYSIS OF THE MEMBRANE COMPONENTS

4.1. Light Absorption

4.1.1. Chromoproteins in Biomembranes

The quantitative analysis of colored material by light absorption is based on Lambert–Beer's law, but this principle may not be applicable to the chromo-proteins in the mitochondrial membrane because of its particulate nature.

The effect of the turbidity of the membrane suspension on the light absorption can partly be avoided by the use of an integral sphere, opal glass, end-on type photomultiplier or dual wavelength spectrometer. The suspension may be clarified by the addition of detergents. Isolation of prosthetic groups such as acid extraction of flavin or HCl–acetone extraction of hemes a and b is the more accurate method.

4.1.2. Extinction Coefficients of Chromoproteins

In order to estimate the amounts of flavoproteins and cytochromes in the mitochondrial membrane, extinction coefficients of these components must be known. However, intrinsic proteins such as cytochromes b_T, b_K, a, and a_3 have never been purified. Moreover, overlapping of the absorption spectra of these components and changes of the extinction coefficients in the membrane must be considered. With these reservations, several workers adopted tentative extinction coefficients.

For example, Estabrook and Holowinsky (1961) reported the millimolar extinction coefficients of the cytochromes in oxidoreduction difference spectra. These are

$$
\begin{aligned}
a_3 &= 90 \quad (\text{at } 444 - 455 \text{ nm}) \\
a &= 16 \quad (\text{at } 605 - 625 \text{ nm}) \\
b &= 20 \quad (\text{at } 562 - 575 \text{ nm}) \\
c + c_1 &= 19.1 \ (\text{at } 551 - 540 \text{ nm}) \\
\text{flavin} &= 11.5 \ (\text{at } 465 - 500 \text{ nm})
\end{aligned}
$$

The relative value or percent reduction of these components can be obtained with less difficulty (Chance et al., 1970).

4.2. Gel Electrophoresis of Membrane Components

4.2.1. Electrophoresis of Proteins in the Presence of Detergents

Owing to the insoluble nature of membrane proteins, separation of these components must be performed in the presence of strong detergents such as dodecylsulfate. Dodecylsulfate plus reducing agents dissociates membrane proteins into subunits and forms rod-shaped complexes composed of dodecylsulfate (1.4 g/g protein) and the protein (or polypeptide subunit). Since the diameters of the rod-shaped complexes are constant irrespective of their molecular weights, the lengths of the complexes are determined by the molecular weights of the proteins (Weber and Osborn, 1969). Moreover, the charge of the complex is mainly determined by the bound dodecylsulfate. For these reasons, electrophoretic mobilities of proteins in the presence of dodecylsulfate are mainly determined by their molecular weights.

4.2.2. Methods of Dodecylsulfate Electrophoresis

4.2.2a. The Buffer and the Gel. Electrophoresis of membrane proteins is performed in 10% polyacrylamide gel at 0.1% sodium dodecylsulfate (Weber and Osborn, 1969). In a final volume of 1 liter, 7.8 g $NaH_2PO_4 \cdot H_2O$, 38.6 g $Na_2HPO_4 \cdot 7H_2O$ (or 20.5 g Na_2HPO_4), and 2 g sodium dodecylsulfate are dissolved in water at room temperature. In a final volume of 100 ml, 22.2 g acrylamide and 0.6 g methylene-bis-acrylamide are dissolved in water and kept at 4 °C in the dark.

Twelve glass tubes (5 by 90 mm) are washed, dried, and stoppered at the bottoms on a special stand. In a beaker, 11.2 ml buffer, 10 ml acrylamide solution, 1.0 ml freshly made 1.5% ammonium persulfate, and 30 μl tetraethylene diamine are mixed in this order. The glass tubes are filled with the mixture up to the 8-cm mark with a Pasteur pipette, and the tops are filled with water and allowed to stand for 2 hr to form polyacrylamide gel.

4.2.2b. Dissociation and Electrophoresis of the Proteins. In a final volume of 30 μl, 10–80 μg membrane proteins, 5 μl of 5% sodium dodecylsulfate, 5 μl 2-mercaptoethanol, 5 μl of 0.05% bromphenol blue, and 5 μl of 50% glycerol are added and incubated at 37 °C for 2 hr. The water on the top of the gels in the glass tubes is removed. The glass tubes are placed in the gel rack, and the upper and lower reservoirs are filled with the buffer diluted with an equal volume of water. The protein samples are put on the top of the gels with a microsyringe, and an electric current of 7 mA per tube is applied (anode at the lower reservoir). The voltage should be about 50 V at room temperature. After 5 hr, the blue band of bromphenol blue approaches the bottom, and the current is stopped.

4.2.2c. Staining the Gel. The gels are removed from the tubes with needles and put in separate test tubes containing 2% TCA–20% methanol solution for a few hours. Then the gels are stained for a few hours at 25 °C in 10 ml of mixture containing 25 mg Coommassie blue, 9 ml of 50% methanol, and 1 ml acetic acid. The staining solution is poured off, and the gels are rinsed with water and destained at 37 °C overnight in a solution containing 5% methanol–70% acetic acid, with one change of the solution.

4.2.3. Collection of the Pure Membrane Proteins

About 1 mg of the membrane proteins is fractionated on the 12 gels as described above. One gel is stained to detect the protein band, and the others are corrected for differences in mobilities by the relative positions of the bromphenol blue. Each gel is cut at the desired position, and the 11 cylindrical gel fragments containing the same band are collected and partially lyophilized for 2 hr to make the gel small enough to put in the glass tube again.

Newly prepared gel mixture is filled up to the 2-cm mark of the glass tubes and polymerized as described above. The six lyophilized fragments are layered on the new gel in one tube and swollen again by addition of twofold diluted buffer. After the gels are swollen, cellophane tubing (5 cm long) is tightly fixed to the bottoms of the glass tubes, 0.3 ml of the twofold diluted buffer is put in the tubing with exclusion of air bubbles, and the ends are tied. The glass tubes are put on the racks, and electrophoresis is carried out as described before but for 18 hr.

The cellophane bag contains the band protein, which can then be used as antigen or for other chemical purposes. If reconstitution of activity after removal of the detergent proves to be possible (Weber and Kuter, 1971), this method of purification may be more useful.

4.3. Immunological Localization of Membrane Proteins

4.3.1. Antibody as a Tool of Membrane Biochemistry

The antigen–antibody reaction is highly specific and takes place only when the large antibody is accessible to the antigen on the membrane. The localization of the antigen in the membrane can thus be elucidated.

4.3.2. Preparation of Antibody

One milliliter of an antigen (2–10 mg/rabbit) is slowly mixed with 1 ml Freund's complete adjuvant (Difco, Michigan) and sonicated for 5 min in a Branson sonifier at room temperature. The sonicate is injected in both sides of the back and in the footpads of two rabbits (body weight 2–3 kg). After 4 weeks, the antigen (5 mg/rabbit) is dissolved in 1 ml of 0.9% NaCl and injected into their ear veins. This boosting is repeated a few times. Five days after the final boosting, 40 ml blood is taken from their ear veins. The blood is allowed to clot at room temperature for 5 hr, then is centrifuged at 15,000 $\times$ g for 10 min and filtered through a Millipore filter of pore size 0.45 μ. The serum is kept frozen at $-70\,°C$ or lyophilized.

4.3.3. Localization of an Antigen by the Antibody

The example of experimentation with anti-OSCP is presented. As shown in

Table V. Effect of Anti-OSCP Serum on Submitochondrial Particles and Alkali-Treated Particles[a]

Particles	Additions in this order	^{32}Pi-ATP exchange (nmoles/mg/min)
SMP[b]	Control serum	54.3
SMP	Anti-OSCP serum	55.9
A particles[c]	OSCP + F$_1$, control serum	72.8
A particles	OSCP + F$_1$, anti-OSCP serum	39.5
A particles	Control serum, OSCP + F$_1$	58.7
A particles	Anti-OSCP serum, OSCP + F$_1$	13.3

[a] To 250 μg particles, 0.1 ml serum (see Section 4.3.2.), 40 μg F$_1$ (see Section 5.1.1.), 2 μg OSCP (see Section 5.1.2), and other components as described in Section 7.1.4 were added. In the case of submitochondrial particles, only serum was added and incubated at room temperature for 20 min. In the case of A particles, the serum or F$_1$ plus OSCP was added first, as indicated in the table, and after 10 min F$_1$ plus OSCP or serum was added. After 10 min, the mixtures were assayed for ^{32}Pi-ATP exchange reaction as described in Section 4.5.2.
[b] Submitochondrial particles prepared as described in Section 6.1.4.
[c] Alkali-treated particles prepared as described in Section 6.1.2, at pH 9.2.

Table V if anti-OSCP is added before F$_1$, the energy transfer activity is inhibited. However, if F$_1$ is added before anti-OSCP, the antibody is not effective. In the case of submitochondrial particles, in which both F$_1$ and OSCP are present, anti-OSCP is not effective.

4.4. Radioimmunoassay of the Membrane Proteins

4.4.1. Quantitative Assay of Energy Transfer Proteins

The proteins of energy transfer have no specific light absorptions and often have no enzyme activity. The enzyme activity, if present, is not proportional to the amount of the protein, owing to the presence of inhibitors or to the environment in the membrane. Radioimmunoassay is one possible method to estimate the amount of the protein on the surface of the membrane.

4.4.2. ^{3}H-acetyl-F$_1$

To estimate the amount of F$_1$ by radioimmunoassay, radioactive F$_1$ is necessary. ^{3}H-Acetyl F$_1$ is prepared as described in Kagawa (1967b).

One and one-half milliliters of solution of F$_1$ (25 mg) is prepared as described in Section 7.1 and is passed through a Sephadex G50 column equilibrated with 10 mM NaHCO$_3$, 10 mM sodium acetate, 1 mM ATP (pH 7.4) at 25°C. The fraction eluted at K_d 0 is freed from NH$_4^+$, and this fraction is mixed with 5 μl of 2.5% ^{3}H-acetic anhydride (1 mCi/mg) dissolved in dry acetone at 25 °C. After 10 min, the labeled F_1 is passed through the new column, which is prepared as described above, and precipitated from the eluate by the addition of an equal volume of saturated ammonium sulfate.

4.4.3. Radioimmunoassay of F_1

4.4.3a. Step 1. Interaction of Anti-F_1 and the Membrane. The anti-F_1 should be a precipitating antibody. For this purpose, anti-F_1 WRA mouse serum is used. The membranes (0.5–2 mg) such as submitochondrial particles or ASU particles are incubated with 0.1 ml of the serum or the control serum in a final volume of 1 ml at 25 °C for 10 min, and the mixture is centrifuged at 105,000 × g for 20 min at 20 °C.

4.4.3b. Step 2. The Interaction of the Remaining Anti-F_1 and ^{3}H-acetyl F_1. The resulting supernatant is mixed with various amounts of ^{3}H-acetyl F_1, which is dissolved at a protein concentration of 1 mg/ml in the solution described in Section 7.1, at 25 °C for 20 min. The mixture is centrifuged in a Coleman centrifuge at 14,000 rpm in a small plastic tube (diameter 3 mm), and the washed precipitate is counted for radioactivity. The titration of the anti-F_1 serum with the labeled F_1 usually gives a ratio of 3 μg F_1 per 1 μl serum. This method reveals the amount of F_1 removed from submitochondrial particles during the preparation of the deficient particles (Table VI).

Table VI. Radioimmunoassay of F_1 in the Particles[a]

Components in step 1		^{3}H-Acetyl F_1 ppt. in step 2 (μg)
None		0.02
Control serum	2 μl	0.00
Anti-F_1 serum	1 μl	4.00
Anti-F_1 serum	2 μl	7.77
Anti-F_1 serum plus SMP[b]	2 μl	1.55
Anti-F_1 serum plus ASU particles[c]	2 μl	3.69

[a] The radioimmunoassay of F_1 was performed as described in Section 4.4.3. The ratio of serum to the particles was 0.1 ml per 1 mg particle protein. The specific ATPase activity and radioactivity of ^{3}H-acetyl F_1 used in this experiment were 81.5 μmoles Pi/mg/min and 25,000 cpm/mg, respectively.
[b] Submitochondrial particles prepared as described in Section 6.1.4.
[c] Alkali–Sephadex–urea treated particles prepared as described in Section 6.1.3.

4.5. Enzyme Assay of the Membrane System

4.5.1. Assay of ATPase

The activity is measured with the particles (50–500 μg) or pure F_1 (1–2 μg) in the presence and absence of 5 μg rutamycin, with 5 mM ATP, 5 mM $MgSO_4$, 20 mM potassium phosphoenolpyruvate, 32 μg pyruvate kinase [ATP : pyruvate phosphotransferase, E.C. 2.7.1.40], and 20 mM tris SO_4 (pH 7.4) in a final volume of 1.0 ml at 30 °C for 10 min (Kagawa, 1967a). The reaction is stopped by the addition of 0.1 ml of 50 % TCA. After centrifugation at 2000 rpm in a clinical centrifuge for 5 min, 0.5 ml of the supernatant is mixed with 3.25 ml water, 1 ml of 2.5 % ammonium molybdate in 5 N H_2SO_4, and 0.25 ml aminonaphthol sulfonate solution. (The solution is prepared by dissolving 0.5 g aminonaphthol sulfonic acid in 195 ml of 15 % $NaHSO_3$ and 5 ml of 20 % Na_2SO_3). The mixture is incubated at 30 °C for 10 min, and the light absorption of the mixture is measured at 660 nm. Oligomycin (or rutamycin) sensitivity is expressed as percent inhibition by the inhibitor.

4.5.2. Assay of Pi-ATP Exchange

The reconstituted vesicles (see Section 7.2) or deficient particles (see Section 6.1) are mixed with coupling factors as described in Section 7.1. For 500 μg of these particles, distilled water is added to obtain a final volume of 0.5 ml, 0.5 ml Pi-ATP exchange stock solution is added, and the mixture is incubated for 10 min at 30 °C (Kagawa and Racker, 1971). This stock solution (0.5 ml) contains 20 μmoles Pi (pH 7.4), 100,000 cpm ^{32}Pi, and 10 μmoles Mg-ATP (pH 7.4). After the incubation, the mixture is combined with 0.1 ml of 50 % TCA and centrifuged at 2000 rpm in the clinical centrifuge for 5 min, and 0.5 ml of the supernatant is pipetted into 5 ml of 1 % ammonium molybdate in 0.25 N $HClO_4$. The resulting yellow mixture is extracted twice with 5 ml isobutanol–benzene (1:1) solution (saturated with water) for 1 min on a Vortex mixer. One milliliter of the lower layer is placed on a planchet, dried, and counted in a gas flow counter for 1 min. The counts are converted into nmoles/mg/min by specific radioactivity of Pi, zero time blank, and the particle protein, which is determined by Lowry's method (1951).

4.5.3. Assay of Oxidative Phosphorylation

The particles (250–500 μg) pretreated as described in Section 7.1 are quantitatively transferred to the chamber of a Clark-type oxygraph (polarographic electrode covered with O_2-permeable membrane) with a polarization voltage of -0.6 V (Estabrook and Pullman, 1967). The oxygen uptake is measured in a final volume of 1.0 ml, which is made up with water containing 50 units hexokinase, 50 μmoles glucose, 0.2 μmole EDTA, 5 μmoles ATP, 5 μmoles $MgSO_4$, ^{32}Pi (100,000 cpm), 20 μmoles Pi buffer (pH 7.4), and substrates such as 10 μmoles sodium succinate. In the case of third-site phosphorylation, 20 μmoles freshly neutralized sodium ascorbate, 0.5 μmole phenazine-methosulfate, and 2 μg antimycin A are added to the chamber. In this case, the addition of external cytochrome c inhibits phosphorylation of sub-mitochondrial or reconstituted vesicles. In the case of NADH oxidation, 0.5 μmole NAD, 10 μmoles ethanol, 50 μg alcohol dehydrogenase [alcohol : NAD oxidoreductase, E.C. 1.1.1.1], and 0.2 unit aldehyde dehydrogenase [aldehyde : NAD oxidoreductase, E.C. 1.2.1.3] are added to regenerate NADH. The oxygen uptake is measured on the recorder at 25 °C, and the reaction is stopped by the addition of 0.1 ml of 50% TCA. When ascorbate or other reducing agent is added to the assay, these are oxidized with periodate before the extraction of Pi.

The ^{32}Pi incorporation into glucose-6-phosphate is determined by extraction of added ^{32}Pi by isobutanol–benzene, as described in the preceding section. The oxygen uptake is calculated from the percent decrease of polarographic current and the value for the oxygen content of air-saturated water (260 nmoles O_2/ml at 25 °C). The P/O ratio is the ratio of μmoles ^{32}P in glucose-6-phosphate to μatoms of oxygen consumed during the reaction.

The buffers have a profound effect on oxidative phosphorylation (Christiansen et al., 1969).

4.5.4. ATP-Driven Transhydrogenase

In mitochondrial membrane, the reduction of NADP with NADH requires energy, which is derived from either electron transport or ATP hydrolysis. The reverse reaction can cause transport of ions.

The particles (250–500 μg) are pretreated as described in Section 7.1 and added to a final volume of 3.0 ml containing 6 nmoles NAD, 90 μg alcohol dehydrogenase, 20 μl of 1 mM rotenone (inhibitor of complex I) in 95% ethanol, 0.3 μmole NADP, 300 μg bovine albumin, 150 μmoles tris SO_4 (pH 7.4), and 50 μg pyruvate kinase [ATP : pyruvate phosphotrans-ferase, E.C. 2.7.1.40] at 25 °C. The reaction is initiated by the addition of 3 μmoles ATP and 6 μmoles phosphoenolpyruvate, and fluorescence enhance-ment is measured with a fluorometer (exciting light at 340 nm and emission light at 475 nm), or the optical density increment at 340 nm is recorded with a spectrophotometer (Estabrook and Pullman, 1967).

4.5.5. ATP-Driven NAD Reduction by Succinate

The particles (250–500 μg) are pretreated as described in Section 7.1 and then added to a final volume of 3.0 ml containing 300 μg bovine albumin, 15 μmoles sodium succinate, 6 μmoles $MgSO_4$, 150 μmoles tris SO_4 (pH 7.4), 6 μmoles sodium sulfide, 0.6 μmole NAD, and 60 μg pyruvate kinase at 25°C. The reaction is initiated by the addition of 3 μmoles ATP and 6 μmoles phosphoenolpyruvate. The reduction of NAD is measured by fluorescence (475 nm) excited at 340 nm or by light absorption at 340 nm, as described in the preceding section.

5. PURIFICATION OF EXTRINSIC PROTEINS

5.1. Proteins of Energy Transfer

5.1.1. Purification of Mitochondrial ATPase (F_1)

5.1.1a. ATPase (or Coupling Factor 1, F_1) and Its Activity Units. F_1 is a globular protein of 300,000 daltons (see Section 1.3) and is a coupling factor of oxidative phosphorylation as well as the ATPase of mitochondria (Fig. 1G).

One unit of ATPase activity is defined as that amount of F_1 which hydrolyzes 1 μmole ATP per minute under the conditions described in Section 4.5.1. One unit of coupling factor activity is defined as that amount of F_1 which results in 50% restoration of ^{32}Pi-ATP exchange activity of A particles (500 μg) in the presence of 10 μg OSCP under the conditions described in Section 7.1.

All procedures except step 1 are performed at room temperature (25°C) to prevent dissociation of F_1 in the cold unless F_1 is in the state of ammonium sulfate precipitate.

5.1.1b. Step 1. Freezing-Thawing and Mild Sonication to Remove Soluble Proteins. Light-layer beef-heart mitochondria (Blair, 1967) (12 g, 60 mg/ml), kept frozen at -70°C, are thawed and diluted with an equal volume (200 ml) of 0.25 M sucrose and centrifuged at 20,000 × g for 20 min. The pellet is homogenized in 300 ml of 0.15 M sucrose, 0.1 mM $MgCl_2$, and 10 mM tris SO_4 (pH 7.4), and sonicated in 30-ml batches in a Raytheon sonicator (10 kc/sec, 0°C) at maximum output for 30 sec. The suspension is centrifuged at 78,000 × g for 1 hr in a Spinco rotor 30. When step 1 is omitted, the resulting F_1 is less pure and yellow.

5.1.1c. Step 2. Sonication to Detach F_1. The pellet is mixed with 240 ml of 0.25 M sucrose, 1.6 ml of 0.5 M EDTA (*p*H 7.4), and 4.0 ml of 0.2 M ATP (*p*H 7.4), and diluted with distilled water to a final volume of 400 ml. The dilute suspension is divided into 200-ml batches, put in a 600-ml steel beaker, and then sonicated with a Branson sonifier (model J32 with 1.25-inch step horn) at maximum output. The temperature of the suspension is allowed to rise to 45 °C during the first 10 min of sonication. After an interval of several minutes, the suspension is sonicated again for another 10 min, and the temperature rises to 55 °C. The suspension is cooled to 25 °C and centrifuged at 78,000 × *g* for 90 min. The resulting supernatant is saved.

5.1.1d. Step 3. Isoelectric Precipitation. A Sorvall RC2B centrifuge is made ready at room temperature. The *p*H of the supernatant is adjusted to 5.4 (glass electrode *p*H meter) with 1 N acetic acid, which is added dropwise through a Pasteur pipette during stirring. The mixture is centrifuged at 14,000 × *g* for 5 min, and the resulting supernatant is adjusted to *p*H 6.7 with 2 M tris, within 10 min after the addition of acid.

5.1.1e. Step 4. Protamine Precipitation. Two grams of pure protamine sulfate (from Lilly) is dissolved in 400 ml of distilled water.

Pilot Precipitation with Protamine. One milliliter of the *p*H 6.7 supernatant is put in each of four Sorvall glass centrifuge tubes, and 10, 20, 30, or 40 μl of the 0.5 % protamine sulfate is added. These are centrifuged at 18,000 × *g* for 5 min, and 5 μl of each of the resulting supernatants is assayed for ATPase activity (Section 4.5.1) (Penefsky *et al.*, 1960). The amount of protamine solution which is enough to precipitate membrane fragments but allows more than 80 % of F_1 to stay in supernatant is determined.

Large-Scale Protamine Precipitation. To the *p*H 6.7 supernatant, the protamine sulfate solution is added at the rate of 10 ml/min with stirring in of an amount proportional to that found satisfactory in the above pilot experiments. After 5 min, the turbid suspension is centrifuged at 15,000 × *g* for 5 min. The resulting supernatant is mixed with 0.4 volume of the protamine sulfate, and this white, turbid suspension is centrifuged in transparent tubes at 15,000 × *g* for 15 min.

5.1.1f. Step 5. Ammonium Sulfate Fractionation. The resulting small precipitate is dissolved carefully with the help of a glass rod in 10 ml of solution containing 0.25 M sucrose, 0.4 M ammonium sulfate, 10 mM tri. SO_4, and 2 mM EDTA. The solution is centrifuged at 15,000 × *g* for 5 min, and the small amount of precipitate is discarded. To the supernatant, an equal amount of saturated ammonium sulfate (pH adjusted to 7.4 by NH_4OH) is slowly added, and the mixture is cooled in an ice bath. After 10 min, the suspension is centrifuged at 15,000 × g for 10 min. The precipitate is warmed and dissolved in 10 ml of solution (pH 7.4, 25 °C) containing 4 mM ATP, 2 mM EDTA, 10 mM tris SO_4, and 0.25 M sucrose. An equal volume of saturated ammonium sulfate is added as described above.

5.1.1g. Step 6. Heat Treatment. An aliquot (50 μl) of the suspension is centrifuged, and the precipitate is dissolved in 1 ml water; then the protein concentration is determined by Lowry's method (1951). The remaining suspension is centrifuged at 15,000 × g for 10 min and dissolved in the following solution to yield a final protein concentration of 10 mg/ml. The solution (pH 7.4, 25 °C) contains 30 mM ATP, 2 mM EDTA, 0.25 M sucrose, and 10 mM tris SO_4. The dissolved F_1 solution is put in 100-ml cylinders, and the temperature is brought to 64 °C with constant stirring with a thermometer in a 72 °C bath, and then quickly transferred to a 64 °C bath. After 4 min heating, the solution is cooled to room temperature and centrifuged at 15,000 × g for 10 min at 25 °C and the supernatant is saved.

5.1.1h. Step 7. Second Ammonium Sulfate Fractionation. An equal volume of saturated ammonium sulfate solution is added to the supernatant, and then the process of centrifugation, dissolving, and reprecipitation is repeated as described in step 5 except that the dissolving solution is 0.25 M sucrose, 10 mM tris SO_4 (pH 7.4), and 2 mM EDTA (pH 7.4). The final F_1 suspension in 50% saturated ammonium sulfate can be stored for months.

5.1.1i. Comments on the Preparation. For smaller-scale preparations (about 1 g of mitochondria), step 2 can be performed in a Raytheon sonication (10 kc/sec) for 30 min, with the circulating water at 40 °C (the internal temperature becomes 55 °C at the end).

  F_1 with latent ATPase activity can be obtained from the step 3 supernatant by fractionation on a DEAE-cellulose column (see Section 3.4) at pH 7.8 at room temperature. The fraction containing F_1 is eluted between 70 and 110 mM NaCl at pH 7.8. The heating step should be avoided to prevent the release of ATPase inhibitor from F_1.

5.1.1j. The Yield, the Purity, and the Specific Activity of F_1. The yield of F_1 activity from the crude extract is 300%, because of the removal of ATPase inhibitor during the heating step. The yield of protein from the extract is 1–2%. Considerable amounts (over 60%) of F_1 still remain on the membrane after the sonication.

The protein is homogeneous (Fig. 1G) or even partially crystallized (Fig. 1H). The specific activity of ATPase is about 100 units/mg and that of coupling factor is about 40 units/mg. These two activities of F_1 protein are consistent with the F_1 content of submitochondrial particles (10% of total proteins). Although ATPase activity may be masked with ATPase inhibitor, the activity can be fully obtained in AS particles which show ATPase activity of 10 units/mg.

5.1.2. Purification of OSCP

5.1.2a. Oligomycin Sensitivity Conferring Protein (OSCP) and Its Activity Unit. OSCP is a basic protein of 18,000 daltons, which specifically connects F_1 to the mitochondrial membrane, restores phosphorylation of A particles with F_1, and renders F_1 sensitive to oligomycin (see Section 1.3).

One unit of OSCP is defined as the amount of OSCP which results in 50% reactivation of ^{32}P-ATP exchange activity of 500 μg A particles prepared at pH 9.5 (see Section 6.1) in the presence of 40 μg F_1 in the conditions specified in Section 7.1.

The activity can also be assayed with 100 μg TUA particles (see Section 6.1), which can confer oligomycin sensitivity on 2 μg F_1 in the presence of OSCP (Kagawa and Racker, 1966a; Bulos and Racker, 1968; MacLennan and Tzagoloff, 1968).

5.1.2b. Step 1. Preparation of F_4 (Conover et al., 1963). Ten grams of light-layer beef-heart mitochondria (Blair, 1967) is homogenized in 200 ml of 0.3 M KCl and 1 mM EDTA and is mixed with 100 ml cold 1.2 M ammonium hydroxide for 10 min at 0 °C. The mixture is centrifuged at 105,000 × g for 1 hr. The clear supernatant is carefully separated from the fluffy precipitate and adjusted to pH 7.4 with 10 N acetic acid, then centrifuged again at 105,000 × g for 1 hr to remove the precipitate.

5.1.2c. Step 2. Protamine Treatment. The supernatant (F_4) obtained in step 1 (2.04 g, 330 ml, 6.2 mg/ml) is mixed with a few milliliters of 0.1 M tris to bring the pH to 7.4 and stirred with 63 ml of 1% protamine sulfate (Lilly, unneutralized). After 10 min, it is centrifuged at 10,000 × g for 10 min and the bulky gray precipitate is discarded.

5.1.2d. Step 3. Ammonium Sulfate Fractionation. The resulting supernatant is stirred with a $^1/_2$ vol (162.5 ml) cold saturated ammonium sulfate (pH 7.2 with NH$_4$OH). The mixture is centrifuged at 10,000 × g for 10 min in polycarbonate centrifuge tubes so that the resulting very small amount of precipitate can be seen. This precipitate is dissolved with 20 mM tris acetate–1 mM EDTA (pH 8.0) in a final volume of 18 ml.

5.1.2e. Step 4. Carboxymethyl (CM) Cellulose Column Chromatography. CM-cellulose (Whatman CM52) is washed with 50 ml of 2 mM tris acetate (pH 8.0), and the slurry is packed in a short column (2.5 by 2.0 cm). The solution obtained in step 3 (18 ml) is diluted with distilled water to 180 ml. The resulting cloudy solution is applied directly onto the CM-cellulose column. The flow rate of the column becomes very slow, but the centrifugation of the cloudy solution is avoided, since OSCP is aggregated whenever the salt and alkali concentrations are reduced.

The column is washed with 200 ml of 2 mM tris acetate (pH 8.0). This washing may take one whole night. Then the column is washed with 50 ml of 20 mM KCl in 2 mM tris acetate (pH 8.0) to remove unrelated proteins. After this, the column is eluted with 200 ml of 20 mM tris acetate (pH 8.0) containing a linear gradient from 5 mM to 50 mM KCl. During the elution, the flow rate is increased to 20 ml/30 min. The fractions (20 ml) are collected automatically, and 50 μl of the eluate is assayed for ^{32}Pi-ATP exchange activity in the presence of A particles, as described in Section 7.1. The activity to restore the exchange is eluted from the column at 0.20 M KCl. The protein is assayed by Lowry's method (1951).

5.1.2f. Step 5. Concentration of OSCP. The fractions which show high specific activity are dialyzed against 25 vol saturated ammonium sulfate (pH 7.2) for 18 hr. The turbid suspension is centrifuged at 15,000 × g for 10 min, and the precipitate is dissolved in 1 ml of 20 mM tris SO$_4$ (pH 8.0) (final protein concentration 1–5 mg/ml).

5.1.2g. Step 6. Sephadex G100 Column Chromatography. Sephadex G100 is equilibrated with a buffer containing 0.3 M KCl and 20 mM tris HCl (pH 8.8) and packed in a long column (1.0 by 20 cm). After the OSCP solution (step 5) is applied, the column is eluted with the same buffer. The active fraction appears at K_{av} 0.45.

5.1.2h. The Yield, the Purity, and the Specific Activity. The activity yield from F_4 is about 40%. The preparation shows a single band on sodium dodecylsulfate gel electrophoresis. Since about 250 μg F_4 is required to restore 50% [32] Pi-ATP exchange activity of 500 μg A particles (Conover et al., 1963), and only 2.5 μg OSCP is enough in the same condition (Kagawa, 1972a), the purification is a hundredfold. Comparing the molecular weights, one or two molecules of OSCP to each F_1 molecule are required to restore both oligomycin sensitivity of F_1 and oxidative phosphorylation.

5.1.3. Purification of ATPase Inhibitor

5.1.3a. ATPase Inhibitor and Its Activity Unit. ATPase inhibitor is a protein of 5570 daltons, and it is specifically bound to F_1 and inhibits its ATPase activity. One unit of activity is defined as the amount of the inhibitor which results in 50% inhibition of 0.2 unit of ATPase activity of F_1 (see Section 5.1.1) in the conditions specified in Section 7.1. Since 0.2 unit of ATPase corresponds to 2 μg pure F_1, 1 unit of the inhibitor is equivalent to 1 μg F_1. The preparative method is the modification of Horstman and Racker (1970). The temperature during the preparation is 0–4 °C. Centrifugation after step 3 is performed in a heavy-wall glass test tube (12-ml Sorvall).

5.1.3b. Step 1. Alkaline Extraction. Light-layer beef-heart mitochondria (Blair, 1967) are stored at -70 °C at a protein concentration of 50 mg/ml, in 0.25 M sucrose. The mitochondria (400 ml, 20 g protein) are thawed with 5 ml of 1 M tris SO_4 (pH 7.4), 4 ml of 0.5 M EDTA (pH 7.4), and 591 ml of 0.25 M sucrose (final volume 1 liter). To this mixture, 50 ml cold 1 N KOH is added with vigorous stirring, and after 1 min, 10 ml of 5 N acetic acid is added, then the pH (initially usually about 11.5) of the mixture is adjusted to 7.4 by the addition of the acid or alkali within a few minutes. The mixture is centrifuged at 10,000 × g for 10 min to remove the precipitate.

5.1.3c. Step 2. Ammonium Sulfate Fractionation. The supernatant (900 ml) is stirred with 270 g ammonium sulfate, and after several minutes the mixture is centrifuged at 10,000 × g for 10 min. To the supernatant, 20 g/100 ml ammonium sulfate is added, and the mixture is centrifuged at 15,000 × g for 10 min. The precipitate is dissolved in 140 ml of 0.25 M sucrose.

5.1.3d. Step 3. TCA Fractionation. To the solution obtained in step 2, 12 ml of 50% TCA is added and the mixture is centrifuged at 15,000 × g for 5 min. The precipitate is suspended in 10 ml water and homogenized in a Teflon homogenizer. The mixture is adjusted to pH 5.0 with stirring and centrifuged at 15,000 × g for 5 min to remove the precipitate.

5.1.3e. Step 4. Ethanol Fractionation. The supernatant is adjusted to pH 7.4 with 1 N KOH. To 1 vol of the solution, 0.25 vol of saturated ammonium sulfate is added, then 2 vol of 95% ice-cold ethanol, and the mixture is centrifuged at 15,000 × g for 10 min. The precipitate is homogenized in a minimum volume of 10 mM tris SO_4 and centrifuged at 15,000 × g for 10 min to remove the precipitate. The supernatant is treated with ammonium sulfate and ethanol as before. The wall of centrifuge tube is dried with a stream of N_2 to remove the ethanol, and the precipitate is dissolved as before.

5.1.3f. Step 5. Heat Fractionation. The precipitate dissolved in the centrifuge tube is heated at 90 °C for 3 min with stirring, cooled, and then centrifuged at 15,000 × g for 5 min to remove the precipitate.

5.1.3g. Comments. Step 1 can be replaced with pyridine extraction (Nelson *et al*, 1973). DEAE-cellulose fractionation with 5 mM ammonium sulfate (pH 7.4) is possible after step 4.

5.1.3h. The Yield, the Purity, and the Specific Activity. The activity yield from the alkaline extract by this procedure is about 20%, and the purification is about fortyfold. As discussed in Section 5.1.3a, 1 unit corresponds to 1 μg F_1 (300,000 mol wt). In fact, the highest specific activity of single-component ATPase inhibitor dimer (12,000 mol wt) is 25,000 units/mg.

5.2. Proteins of Electron Transport

5.2.1. Summary of Purification of Cytochrome c

5.2.1a. Cytochrome c and Its Properties. Cytochrome c is a basic monomeric globular protein of 13,370 daltons (bovine heart), with 1 mole of heme c. Its detailed conformation has been established by X-ray crystallography (Dickerson *et al.*, 1971). Since cytochrome c is commercially obtained, the preparative method is discussed only from the standpoint of membrane biology.

5.2.1b. Step 1. Acid Extraction. Cytochrome c is extracted from beef-heart mince with acids such as TCA (Keilin and Hartree, 1937) and acetic acid (Okunuki, 1960). Since cytochrome c is on the outer surface of mitochondria (see Section 1.5) and easily lost by washing in salt solution, it is usually prepared from intact mitochondria. The surface of the molecule is covered with patches of basic and acidic amino acid groups, and its ionic interaction with negatively charged inner membrane is disrupted by acid. The compact, stable structure of cytochrome c allows a drastic purification procedure at room temperature, which denatures other proteins.

5.2.1c. Step 2. Elution from Ion-Exchange Resin. Since cytochrome c is a basic protein, it is adsorbed on acidic ion-exchange resins such as Amberlite-IRC50 or -XE64 (H^+ form) at pH 6.4 in the presence of dilute (50 mM) ammonium phosphate. The dark red band on the resin can be eluted with 0.5 N ammonium phosphate at pH 7.0. This procedure is repeated.

5.2.1d. Crystallization. The cytochrome c is crystallized from the eluate (protein concentration 4–8%) by the addition of powdered ammonium sulfate (0.4505 g/ml) at pH 6.0–6.5 at room temperature within few days.

5.2.2. Summary of Purification of Succinate Dehydrogenase

5.2.2a. Succinate Dehydrogenase and Its Properties. Succinate dehydrogenase is a protein of 97,000 daltons and contains 1 mole of covalently bound flavin, 7–8 g-atoms of iron, and 7–8 moles of acid-labile sulfide per mole (Davis and Hatefi, 1971). The enzyme is composed of two subunits; one is a flavin–iron protein of 70,000 daltons with 1 mole flavin, 4 irons, and 4 labile sulfides per mole, and the other is iron protein of 27,000 daltons (Hanstein *et al.*, 1971a). Although this enzyme can be isolated by several methods such as *n*-butanol treatment or alkali extraction (Singer, 1966), the chaotropic extraction seems satisfactory (Davis and Hatefi, 1971).

5.2.2b. Chaotropic Extraction of the Dehydrogenase from Complex II. Complex II (Hatefi *et al.*, 1962) is suspended at a protein concentration of about 1% in 50 mM tris HCl (pH 8.0) containing 20 mM sodium succinate and 5 mM dithiothreitol. $NaClO_4$ (8 M) is then added to a final concentration of 0.4 M, and the mixture is allowed to stand at 0 °C. After 20 min, the mixture is centrifuged at 49,000 rpm in a Spinco 50 rotor for 80 min.

5.2.2c. Ammonium Sulfate Fractionation. The resulting supernatant is fractionated with neutralized ammonium sulfate, and the fraction precipitating between 36 and 50% salt saturation is collected after centrifugation at 40,000 rpm for 10 min. The precipitate is immediately frozen in liquid N_2.

5.2.2d. Yield of Succinate Dehydrogenase. When complex II is extracted once again, the yield of purified succinate dehydrogenase is 55–60 % of the total flavin of complex II. The activity is determined with phenazinemethosulfate as an electron acceptor.

6. PURIFICATION OF INTRINSIC PROTEINS

6.1. Deficient Particles

6.1.1. Definition of the Deficient Particles

Deficient particles are submitochondrial particles from which coupling factors are totally or partially removed (Fig. 1D,E). Coupling factors are extractable proteins of the mitochondrial inner membrane and are capable of restoring oxidative phosphorylation. in appropriate deficient particles (see Section 1.2). These factors so far purified are extrinsic proteins. Deficient particles contain all the intrinsic proteins, including electron transport systems and phospholipids, and have vesicular membrane structure. The submitochondrial particles deficient in electron transport components such as cytochrome c, succinate dehydrogenase, and coenzyme can be prepared, but they are not considered deficient particles.

6.1.2. Alkali-Treated Particles (A Particles)

A particles are submitochondrial particles prepared from mitochondria by sonic oscillation at pH 9.2–9.5 in the presence of dilute ammonia (Fig. 1D). They are deficient in both F_1 and OSCP, since mechanical treatment removes F_1 and alkali treatment removes OSCP.

Heavy-layer beef-heart mitochondria are prepared as described by Blair (1967) and are kept frozen at $-70\,^{\circ}C$ for months. Ten grams of the mitochondria is thawed and homogenized in a final volume of 500 ml containing 0.5 ml of 0.5 M EDTA (pH 7.4) and 100 ml of 0.25 M sucrose. The pH of the homogenate (in 100-ml batches) is adjusted to 9.2 or 9.5 with 1 N ammonium hydroxide. The homogenate (in 25-ml batches) is exposed to sonic oscillation for 2 min in a Raytheon sonicator (250 W, 10 kc/sec) with circulating ice-cold water. The sonicated mixture is combined and centrifuged at 26,000 × g for 10 min at 0 °C. The resulting precipitate is discarded, and the supernatant is centrifuged in a Spinco 30 rotor at 30,000 rpm for 90 min. The precipitate is homogenized in 300 ml of 0.25 M sucrose–1 mM EDTA (pH 7.4) and centrifuged again as described above. The final precipitate is homogenized in the same solution at a protein concentration of 30–50 mg/ml and stored at $-70\,^{\circ}C$ in small aliquots (0.2–0.5 ml) to avoid repeated freezing and thawing during the assay and storage.

The yield of A particles is about 30% of the starting mitochondria. Pi-ATP exchange activity of A particles in the presence of a saturated amount of F_1 and OSCP (described in Section 7.1.4) is 70 nmoles/mg/min for the pH 9.2 preparation, and 45 nmoles/mg/min for the pH 9.5 preparation. Although energy transfer reactions of A particles are absolutely dependent on the addition of both F_1 and OSCP, A particles, even in the pH 9.5 preparation, still contain ATPase activity (1 μmole/mg/min) which is oligomycin sensitive; i.e., they contain both F_1 and OSCP. This phenomenon will be discussed in Section 7.1.

6.1.3. Alkali–Sephadex Treated Particles (AS Particles) and Alkali–Sephadex–Urea Treated Particles (ASU Particles)

In order to remove residual F_1 from A particles, urea is used to dissociate F_1 from the deficient particles. Since F_1 is stabilized by the ATPase inhibitor, passage of A particles through a Sephadex column equilibrated with a high concentration of salt is necessary to remove the ATPase inhibitor before the urea treatment.

A large column (5.0 by 50.0 cm) of Sephadex G50 is equilibrated with a solution containing 250 mM KCl, 75 mM sucrose, 30 mM tris SO_4 (pH 8.0), and 2 mM EDTA. A particles are homogenized in the same solution and finally diluted to 40 mg/ml. This homogenate is passed through the column, and the velocity of flow is controlled by watching the brown band and by narrowing the eluting tube so as to pass the band at 10 cm/10 min. The column temperature is 25 $^{\circ}C$. The turbid fractions at the void volume of the column are collected and centrifuged at 30,000 rpm in a Spinco 30 rotor at 0 $^{\circ}C$ for 90 min. The precipitate is homogenized with 0.25 M sucrose and 10 mM tris SO_4 (pH 8.0) at a protein concentration of 50 mg/ml. This preparation is called "AS particles" and shows high ATPase activity (5–8 μmoles P/mg/min).

The AS particles thus obtained are diluted with cold distilled water to a protein concentration of 10 mg/ml. To this suspension, an equal volume of ice-cold freshly prepared 4 M urea containing 4 mM EDTA and 0.1 M tris SO_4 (pH 8.0) is added and mixed. After 1 hr at 0 $^{\circ}C$, the mixture is centrifuged in a Spinco 30 rotor at 30,000 rpm for 1 hr. The resulting precipitate is homogenized with 0.25 M sucrose at a protein concentration of about 40 mg/ml, and the homogenate is centrifuged again. The final precipitate is homogenized with 5 mM DTT, 0.25 M sucrose, and 10 mM tris SO_4 (pH 7.4) at a protein concentration of about 50 mg/ml. This preparation is called "ASU particles" (Fig. 1E), has very little ATPase (0.02 μmoles P/mg/min) and shows very few F_1 molecules on the surface of the vesicles (Racker and Horstman, 1967).

The activities and assay of energy transfer reactions of ASU particles are described in detail in Section 7.1.4.

6.1.4. Submitochondrial Particles

Submitochondrial particles are vesicular inner membranes obtained from mitochondria by sonic disruption at neutral pH (Fig. 1B,C). These particles are not deficient particles and are active in oxidative phosphorylation without the addition of coupling factors. There are many described preparations such as ETPH (*cf.* Estabrook and Pullman, 1967). The method described here is the simplest and the most convenient (Kagawa, 1967*a*).

The frozen beef-heart mitochondria (Blair, 1967) (Fig. 1A) are homogenized with distilled water and pyrophosphate buffer (pH 7.4) to a final protein concentration of 20 mg/ml and final pyrophosphate concentration of 10 mM. The homogenate is exposed to sonic oscillation in 30-ml batches in a Raytheon sonicator (10 kc/min) at 0 °C for 2 min. The sonicated mixture is centrifuged at 26,000 × g for 15 min, and the resulting supernatant is centrifuged at 30,000 rpm in a Spinco 30 rotor for 1 hr. The precipitate is washed with 0.25 M sucrose and homogenized in the same sucrose solution to yield a suspension of 50 mg/ml. The yield of protein from mitochondria is 40%.

This preparation is active in oxidative phosphorylation as well as energy transfer reactions such as H^+ accumulation. The electron micrograph of these particles shows vesicles on which F_1 molecules (90 Å spheres) are fully loaded (Fig. 1B,C).

6.1.5. Trypsin–Urea Treated Particles (TU Particles) and Trypsin–Urea–Alkali Treated Particles (TUA Particles)

6.1.5a. Complete Removal of F_1 and OSCP. ASU particles still contain a considerable amount of OSCP not extracted during the first alkali treatment. More complete removal of F_1 is achieved by the use of trypsin to remove ATPase inhibitor, and then by the use of urea. The TU particles are then extracted with alkali to remove OSCP. The preparative method is essentially the same as that described by Kagawa and Racker (1966*a*) and Bulos and Racker (1968).

6.1.5b. Step 1. Trypsin Digestion. Trypsin (1 mg/ml) is dissolved in 1 mM H_2SO_4, and trypsin inhibitor (2 mg/ml) is dissolved in 10 mM tris SO_4 (*p*H 8.0). The submitochondrial particles (see Section 6.1.4) are diluted to 10 mg/ml with 50 mM tris SO_4 (*p*H 8.0) and 1 mM EDTA. Trypsin (3 μg/mg protein) is added to the mixture, and it is incubated at 30°C for 45 min; then the trypsin inhibitor (15 μg/mg protein) is added to the mixture.

The protein yield from submitochondrial particles is about 80%. These trypsin-treated particles (T particles) have lost ATPase inhibitor and show high ATPase activity (4–8 moles P/mg/min).

6.1.5c. Step 2. Urea Treatment (TU Particles). The trypsin-treated mixture is chilled to 0°C, and an equal volume of ice-cold freshly prepared 4 M urea is added. After incubation at 0°C for 45 min, the mixture is centrifuged at 30,000 rpm in a Spinco 30 rotor for 45 min.

The precipitate is rinsed with a few milliliters of 0.25 M sucrose and homogenized in 0.25 M sucrose to a protein concentration of about 40 mg/ml. The homogenate is centrifuged and homogenized again as described above.

The preparation, called "TU particles", has lost ATPase activity (0.01 μmole P/mg/min) and the 90 Å spheres on the surface of vesicles, and has acquired specific binding capacity of F_1 (Kagawa and Racker, 1966*a,c*). The protein yield from the submitochondrial particles is about 55%. This trypsin–urea treatment can be replaced by chaotropic extraction by I^-, SCN^- (Kagawa and Racker, 1966*c*), or Br^- (MacLennan and Tzagoloff, 1968).

6.1.5d. Step 3. Alkali Extraction (TUA Particles). TU particles are diluted with 0.25 M sucrose and 1 mM EDTA (*p*H 7.4) to a protein concentration of 10 mg/ml at 0°C. The *p*H of this suspension is adjusted to 10.5 with freshly prepared 1 M ammonia. After 10 min, the mixture is sonicated in a Raytheon sonicator (10 kc/sec) for 5 min at 0°C and centrifuged at 40,000 rpm in a Spinco 40 rotor for 1 hr at 0°C. The precipitate is rinsed with a few milliliters of 0.25 M sucrose and homogenized in the cold sucrose to yield a protein concentration of 40 mg/ml. The yield of these particles is about 50% of TU particles.

TUA particles (100 μg) can absorb F_1 (10 μg) and confer oligomycin sensitivity on F_1 (50% inhibition) if saturating amounts of OSCP (1 μg/10 μg TUA particles) are added in a final volume of 0.2 ml containing 1 μmole $MgCl_2$ and 5 μmoles tris SO_4.

6.2. Red–Green Separation

6.2.1. Solubilization of Intrinsic Proteins

The detergent treatment of inner membrane, whether it is submitochondrial particles or deficient particles, results in the separation of a red supernatant which is rich in cytochromes b, c_1, and c and a green precipitate which is rich in cytochromes a and a_3. This first step of membrane fractionation can be achieved by adding the proper concentration of agents such as cholate (2%), deoxycholate (0.5%), Triton X-114 (20%), sodium dodecylsulfate (0.05%), or t-amyl alcohol (10%) at certain concentrations of salt and protein. However, the first two detergents are selected because of their easy removal and protective properties described in Section 2.3.

6.2.2. Separation by Deoxycholate

Separation by deoxycholate, a standard method of membrane fractionation, has been used for isolation of many intrinsic proteins of mitochondria such as the complexes (Wainio, 1970). Submitochondrial particles (23 mg protein/ml) are mixed with 10% deoxycholate (0.3 mg/mg protein, pH 8.0) and solid KCl (final concentration 1 M) at 0°C. The mixture is stirred for 5 min and centrifuged in a Spinco 40 rotor at 40,000 rpm for 20 min. The red supernatant is practically free of cytochrome oxidase.

6.2.3. Separation by Cholate

Cholate (2%) has been used to extract cytochrome oxidase together with cytochromes b and c_1 in the presence of 1.4 M ammonium sulfate (Okunuki, 1960). However, if the ammonium sulfate is lowered to 0.1 M and the centrifugal force is increased, the supernatant is practically free of cytochrome oxidase (Kagawa, 1966b).

Submitochondrial particles (2 g) are mixed with 25 ml of a solution containing 4 mM dithiothreitol, 0.2 mM EDTA, 0.5 M sucrose, 0.4 M ammonium sulfate, and 40 mM tricine-KOH (pH 8.0) at 0°C. The final volume is adjusted to 80 ml with distilled water, and 20 ml of 10% sodium cholate (pH 8.0) is added with stirring at 4°C. After 10 min, the mixture is centrifuged in a 50Ti rotor at 50,000 rpm for 1 hr. The protein concentration of the supernatant is 4 mg/ml.

This supernatant contains intrinsic proteins of energy transfer, which can be measured by ^{32}Pi-ATP exchange activity with F_1 and OSCP after direct dialysis. The conditions of extraction are summarized in Table VII.

Table VII. Conditions of Extraction of Energy Transfer Components from
Submitochondrial Particles with Cholate and Salt[a]

Final concentration during extraction			Extracted proteins (% of SMP)	Extracted phospholipids (μmoles P/mg)	^{32}Pi-ATP exchange of dialyzed extract (nmoles/mg/min)
$(NH_4)_2SO_4$ (mM)	Cholate (%)	SMP[b] (%)			
0	2	2	17.6	2.12	3.58
20	2	2	16.5	2.43	11.95
40	2	2	18.3	2.37	32.55
100	2	2	23.2	2.36	33.90
200	2	2	27.0	2.07	13.45
100	1	2	15.9	2.87	63.75
100	0.5	2	9.1	1.89	4.26
0	2	0.5	19.1	1.89	8.04
100	2	0.5	20.4	2.66	29.90
200	2	0.5	27.6	2.00	7.48
100	1	0.5	18.8	2.34	51.75
100	0.5	0.5	14.9	2.36	16.25

[a] The indicated protein concentration of submitochondrial particles, ammonium sulfate and sodium cholate were incubated at pH 8.0 at 0 °C with the ingredients described in Section 6.2.3 for 30 min, and the mixture was centrifuged at 50,000 rpm in a Spinco 50Ti rotor for 1 hr. The resulting supernatant was dialyzed as described in Section 7.2.3c, and the dialysate was added with F_1 and OSCP as described in Section 7.1.4, then was assayed for ^{32}Pi-ATP exchange reaction as described in Section 4.5.2. The protein was measured by Lowry's method (1951), and phospholipid P was determined by the method described in Section 4.5.1 after ashing with concentrated H_2SO_4–concentrated HNO_3 mixture (1:1) at 220 °C for 1 hr followed by hydrolysis at 100 °C for 10 min.
[b] Submitochondrial particles prepared as described in Section 6.1.4.

The centrifugation can be intensified to 65,000 rpm in a 65 rotor for 2 hr. The optimum sucrose concentration is 0.1 M.

6.3. Proteins of Energy Transfer

6.3.1. Oligomycin-Sensitive ATPase

6.3.1a. Components of Oligomycin-Sensitive ATPase. The isolation and reconstitution of oligomycin-sensitive ATPase revealed that this activity requires F_1, OSCP (or F_4), phospholipids, and some intrinsic proteins (Kagawa and Racker, 1966a,b; MacLennan and Tzagoloff, 1968).

The intrinsic proteins involved in this reaction are not characterized. For example, sodium dodecylsulfate gel electrophoresis of oligomycin-sensitive ATPase prepared by Tzagoloff *et al.* (1968a) revealed at least 14 bands of proteins (Stekhoven, 1972). DCCD-binding protein may be involved (Cattell *et al.*, 1971; Stekhoven, 1972).

6.3.1b. Purification from the Cholate Extract. This preparation is modified from the original purification of oligomycin-sensitive ATPase (Kagawa and Racker, 1966*b*) so as to preserve potential activity of energy transfer such as ^{32}Pi-ATP exchange and H^+ accumulation (Kagawa *et al.*, 1972). The whole procedure is summarized in Fig. 3.

All the procedures are performed at $0\,^\circ C$. The cholate extract described in Section 6.2.3 is mixed with 1/3 vol of saturated ammonium sulfate solution (*p*H 6.4 at twentyfold dilution). After stirring for 10 min, the mixture is centrifuged at 26,000 × *g* for 10 min. The insoluble precipitate is discarded, and the supernatant is mixed with 1/4 vol of the saturated ammonium sulfate

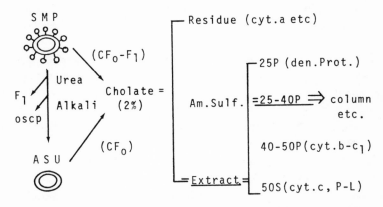

Fig. 3. Fractionation of intrinsic membrane proteins by cholate and ammonium sulfate. SMP, submitochondrial particles prepared as described in Section 6.1.4; ASU, alkali–Sephadex–urea treated particles as described in Section 6.1.3; Residue, precipitate of the red–green separation described in Section 6.2; Extract, supernatant of the red–green separation described in Section 6.2; Am. Sulf., ammonium sulfate fractionation, 25P, precipitate of 25% saturation which contains insoluble denatured proteins; 25–40P, precipitate between 25 and 40% saturation which contains most of the intrinsic proteins of energy transfer; 40–50P, precipitate of 40–50% saturation which contains most of cytochromes *b* and c_1; 50S, supernatant fraction of 50% saturation which contains most of cytochrome *c* and phospholipids; CF_o-F_1, oligomycin-sensitive ATPase depleted in phospholipids which is composed of F_1 and CF_o; CF_o, cholate-treated oligomycin sensitivity conferring factor described in Section 6.3.2. F_1, coupling factor 1 prepared as described in Section 5.1.1.; OSCP, oligomycin sensitivity conferring protein described in Section 5.1.2.

solution. The suspension is centrifuged at 26,000 × g for 20 min. The range of this fractionation corresponds to the ammonium sulfate concentration between 25 and 40% saturation, and the fraction is called "25–40P". The precipitate is dissolved in 1 mM dithiothreitol, 10 mM tricine-KOH (pH 8.0), and 0.25 M sucrose, at a protein concentration of 20–30 mg/ml, and can be kept frozen at $-70\,^\circ$C or lyophilized. The protein yield from submitochondrial particles is 4.8–7.3%. The fraction 25–40P contains 0.1 μmole phospholipid P/mg protein, 0.15 nmole cytochrome b, and practically no cytochrome oxidase. Since 85% of original phospholipids are removed, the oligomycin-sensitive ATPase activity (2.5–5.0 μmoles P/mg/min) is latent unless phospholipid (1 μmole P/mg) is added. If the vesicles are reconstituted from 25–40P as described in Section 7.2, ^{32}Pi-ATP exchange activity of the vesicles is 55–87 nmoles/mg/min. Most of the cytochromes b and c are precipitated between 40 and 50% saturation.

If phospholipase A–treated mitochondria are extracted with less than 2% cholate, a kind of oligomycin-sensitive ATPase called "P_2 fraction" is obtained (Kopazyk et al., 1968) which is partially dependent on phospholipids.

6.3.1c. Purification from the Deoxycholate Extract. This procedure is described by Tzagoloff et al. (1968a), and the preparation contains enough phospholipid to keep its ATPase active, but it is inactive in ^{32}Pi-ATP exchange.

The deoxycholate extract of submitochondrial particles described in Section 6.2 is dialyzed against 8 vol of 10 mM tris acetate (pH 7.5) for 2 hr. The dialyzed red supernatant is centrifuged at 105,000 × g for 15 min. Both dialysis and centrifugation procedures are hard to reproduce, but conditions should be chosen so that 50% of the protein in the red supernatant is recovered in the 15-min supernatant (see Section 3.5).

A 50-ml portion of this supernatant is passed through a column of Sephadex G25 (10 by 25 cm) equilibrated with 10 mM tris acetate (pH 7.5). The turbid supernatant is centrifuged at 105,000 × g for 150 min, and the precipitate is suspended in 0.25 M sucrose, 10 mM tris acetate (pH 7.5), and 1 mM dithiothreitol at a protein concentration of 15 mg/ml and sonicated for 15 sec with a Branson sonifier. The sonicate is added with potassium cholate (0.5 mg/mg protein) and fractionated with saturated ammonium sulfate. The fraction between 25 and 42% saturation is collected and dissolved (25 mg protein/ml) as described in the preceding section, then passed through a Sephadex G25 column equilibrated with 10 mM tris acetate (pH 7.5). The specific ATPase activity is 4–6 μmoles P/mg/min, and the protein yield is 2%.

6.3.2. Intrinsic Proteins of Oligomycin-Sensitive ATPase

6.3.2a. CF_o and Alkali-Treated CF_o. Oligomycin-sensitive ATPase contains some F_1 and OSCP, which are extrinsic proteins. In order to purify the active components in the membrane, F_1 and OSCP must be removed. Instead of submitochondrial particles, ASU particles, TU particles, or TUA particles can be used as starting materials, and procedures similar to these described in Section 6.3.2b are applied to these deficient particles. The preparation called "CF_o" (cholate-treated oligomycin sensitivity conferring factor) is thus obtained from TU particles (Kagawa and Racker, 1966b; Kagawa, 1967a). CF_o contains 0.1 μmole phospholipid P/mg protein. One milligram of CF_o can bind 100 μg ^{3}H-acetyl F_1, and on addition of 1 μmole of phospholipid P, oligomycin-sensitive ATPase is restored.

Alkali-treated CF_o can be obtained from TUA particles with the same procedure. The addition of OSCP is essential to restore the oligomycin sensitivity. Owing to the damaging effect of trypsin on energy transfer reactions, ASU particles instead of alkali-treated CF_o are used as starting material for the reconstitution experiments.

6.3.2b. Fractionation by Detergent–Gel Filtration. Further removal of electron transfer components from oligomycin-sensitive ATPase is attempted by passing the 15-min supernatant of dialyzed deoxycholate extract through a detergent-gradient column (Swanljung, 1971).

The potential energy transfer activity in 25–40P can be preserved after passage through Sepharose 4B or 6B which is equilibrated with a solution containing 1 % cholate, 50 mM ammonium carbonate, 5 mM 2-mercaptoethanol, and ammonia (final pH 10). The fraction corresponding to 300,000 daltons is active, and the activity is stimulated by the addition of the 60,000-dalton fraction (Kagawa, 1972b). It is essential to concentrate each fraction by lyophilization in the presence of 0.2 M sucrose before the reconstitution of active vesicles as described in Section 7.2.

6.3.2c. Dicyclohexylcarbodiimide (DCCD) Binding Protein. One of the intrinsic proteins of oligomycin-sensitive ATPase interacts with an energy transfer inhibitor, N,N'-dicyclohexyl-^{14}C-carbodiimide. The protein has a molecular weight of 10,000 (Cattell et al., 1971; Stekhoven et al., 1972). Beef-heart mitochondria or oligomycin-sensitive ATPase is incubated with the inhibitor (enough to inhibit 95% of ATPase) in the presence of 0.25 M sucrose, 1 mM ATP, 5 mM $MgCl_2$, and tris phosphate buffer (pH 7.6) at 4°C for 16 hr. The treated fragments are washed with 0.25 M sucrose–10 mM tris SO$_4$ buffer (pH 7.6) at 4°C, and the precipitate suspended in the same solution is extracted with 20 vol of a chloroform–methanol (2:1) mixture. The concentrated proteolipids thus extracted are dissolved in a small volume of chloroform–methanol (2:1) and chromatographed on a column of Sephadex LH20, which is equilibrated with chloroform. The column is eluted with chloroform, followed by chloroform containing increasing amounts of methanol. Although the radioactivity is eluted at different methanol concentrations, sodium dodecylsulfate gel electrophoresis of these different fractions reveals an identical band of radioactivity (10,000 daltons). The protein is hard to dissolve and shows no biological activity such as ATPase inhibitor or energy transfer reactions, even when it is reconstituted with other components.

7. REASSEMBLY

7.1. Adsorption of Extrinsic Proteins on the Membrane

7.1.1. Mechanism of Readsorption of Detached Extrinsic Proteins

Theoretically, the formation of a biological structure requires both specific information and net free energy consumption. But neither is added in the reconstitution of oligomers or higher structures from isolated protomers, since the protomers contain the necessary information in their own primary structure and some free energy in a specified solvent.

This self-association process of protomers or simple peptides into higher structure is driven by hydrophobic, electrostatic, disulfide, and hydrogen bonding among the amino acid residues in them (Frieden, 1971). There is evidence to support the idea that the same specific mechanisms are involved in the readsorption of extrinsic proteins on the membrane (Kagawa, 1972a). If the reconstituted structure is as complex as ribosomes (Nomura, 1972), there is a definite order of reconstitution from its components.

7.1.2. General Methods in the Readsorption

The purified extrinsic proteins are adsorbed on the mitochondrial membrane, which is deficient in these proteins when both are mixed in certain conditions summarized in Table VIII. Obviously, the opposite conditions to these which detach the extrinsic proteins are effective: antichaotropic agents for the proteins extracted with a chaotropic agent, divalent cations for those

Table VIII. Methods of Readsorption of the Extrinsic Proteins onto the Deficient Particles

Conditions and Agents	Detachment	Adsorption	Examples
Concentration of proteins	Dilution	Concentration	F_1, OSCP
Special anion	Chaotropic	Antichaotropic	Succinate dehydrogenase
Metal	Chelator (EDTA)	Divalent cation (Mg)	F_1
pH	Alkali (pH 9–10)	Neutral (pH 6–7) or acidic	OSCP, ATPase inhibitor
Salt	High	Low	Cytochrome c
Nucleotides	ATP	ADP-Mg	F_1-ATPase inhibitor
Organic solvent	Add	Remove	
Detergent	Add	Remove	

extracted with a chelator, etc. For example, succinate dehydrogenase is dissociated from complex II by the addition of a chaotropic agent and is readsorbed on it by the addition of antichaotropic agents (Davis and Hatefi, 1972).

7.1.3. Binding of F_1 on the Membrane Deficient in F_1

The membrane preparations deficient in F_1 but not in OSCP, etc., are "Nossal particles" (N particles; Penefsky *et al.*, 1960), TU particles (see Section 6.1), CF_0 (Kagawa and Racker, 1966b), etc. Owing to the presence of ATPase inhibitor in the particles and to the dissociation of F_1 in dilute conditions, the ATPase activity in both precipitate and supernatant of the particles after the addition of F_1 does not represent the quantities of F_1 rebound on the membrane. For example, N particles still contain a large amount of F_1, and the addition of F_1 does not cause an increase of ATPase activity in N particles. For the quantitative assay of binding, the use of ^{3}H-acetyl F_1 (Kagawa, 1967b) is essential.

A suspension of F_1 in 50% saturated ammonium sulfate solution (see Section 5.1) which has been stored at 4°C is centrifuged at 15,000 × g for 10 min. The wall of the centrifuge tube is wiped, and the precipitate is dissolved in a solution (pH 7.4) containing 0.25 M sucrose, 10 mM tris SO_4, 1 mM EDTA, and 2 mM ATP at a protein concentration of 2 mg/ml. This F_1 solution should be kept at room temperature.

F_1-deficient particles (1 mg) are incubated for 10 min at 30°C with 50–100 μg F_1 dissolved in the above-described solution, 4 μmoles $MgCl_2$, and 20 μmoles tris SO_4 (pH 7.4) in a final volume of 0.2 ml. Then the mixture is assayed for ^{32}Pi-ATP exchange, ATPase, etc., as described in Section 4.5.

The binding of ^{3}H-acetyl F_1 (see Section 4.4.2) is specific and takes place only when the membrane has vacant sites for F_1 (Table IX). When F_1 is bound on the membrane, oligomycin sensitivity of F_1 is restored and oxidative phosphorylation of the membrane is recovered.

Table IX. Binding of ^{3}H-Acetyl F_1 onto Mitochondrial Membrane[a]

Mitochondrial membrane	Radioactivity (cpm/2 ml)		ATPase of supernatant (Pi μmoles/min/2 ml)
	supernatant	precipitate	
Mitochondria	26,165	3,550	16.5
S particles[b]	19,793	8,554	24.4
N particles[c]	2,903	24,111	1.61
No particle	26,461	676	21.4

[a] ^{3}H-Acetyl F_1 (61 μg) (see Section 4.4.2) was added to 1 mg of mitochondrial membranes as described in the text, then the mixture was diluted to 2 ml with 10 mM tris SO_4 (pH 7.4) and centrifuged at 150,000 $\times$ g for 30 min.
[b] Submitochondrial particles (see Section 6.1.4).
[c] Nossal particles (see Section 7.1).

7.1.4. Binding of F_1 and OSCP on the Deficient Particles

Mitochondrial membranes deficient in both F_1 and OSCP are A particles, ASU particles, TUA particles (see Section 6), and reconstituted vesicles (described in Section 7.2).

In a test tube, 50 μl of a 5 % solution of defatted bovine serum albumin in 20 mM tris SO_4 (pH 7.4), 5 μl of 0.1 M $MgCl_2$, 20 μl OSCP (200 μg/ml in 0.2 M KCl, 20 mM tris SO_4, pH 8.0), and 20 μl F_1 solution (2 mg/ml as described in the preceding section) are added in this order and incubated at 30°C for 10 min (Kagawa et al., 1973). The mixture is assayed for activities as described in Section 4.5.

The percentage restoration of ^{32}Pi-ATP exchange or P/O ratio is by no means directly proportional to the amount of F_1 or OSCP added to the membranes, but rather is sigmoidal (Kagawa, 1972a). This is the reason why A particles are absolutely dependent on F_1 and OSCP to restore Pi-ATP exchange, although they contain some F_1 and OSCP.

The activities of ASU particles (Racker and Horstman, 1967) that can be restored by the addition of F_1 and OSCP are oxidative phosphorylation, ^{32}Pi-ATP exchange, ATP-driven reduction of NAD by succinate, ATP-driven transhydrogenase, H^+ uptake (Hinkle and Horstman, 1971), fluorescence enhancement of anilinonaphthalene sulfonate (Datta and Penefsky, 1970), and oligomycin-sensitive ATPase. These are all energy transfer reactions, and the requirement for both F_1 and OSCP is summarized in Table X. If the energy is provided by ATP hydrolysis, these reactions are stopped by the addition of oligomycin or other energy transfer inhibitors, while if the energy is derived from oxidoreduction, these reactions are stopped

Table X. Restoration of Energy Transfer Activities in ASU Particles by the Addition of F_1 and OSCP[a]

Energy transfer activities	Units	Activities	
		Controls[b]	Complete system (F_1, OSCP)
1. P/O ratio by succinate oxidation	Ratio	0.0–0.1	0.3–1.1
2. ATP synthesis by succinate oxidation	G6P nmoles/mg/min	8–14	75–180
3. ATP-driven NAD reduction by succinate	NADH nmoles/mg/min	0–10	31–83
4. ATP-driven trans-hydrogenase	NADPH nmoles/mg/min	0–1.2	8–25
5. ^{32}Pi-ATP exchange	AT^{32}P nmoles/mg/min	0–7	33–128
6. Decay of H^+ gradient	Sec	1–2	5–12
7. Respiratory control released by FCCP[c]	Ratio	1.0	3.4
8. H^+ uptake driven by NADH oxidation	H^+ ng ion/mg	1.9–6.2	18.4
9. Fluorescence enhancement of ANS[d]	Arbitrary unit	1.0	15
10. Oligomycin-sensitive ATPase	Pi nmoles/mg/min	10–25	1000–3500

[a] The conditions of binding of F_1 and OSCP onto ASU particles were essentially the same as described in Section 7.1. and activity assays in 1–5 and 10 were as described in Section 4.5 while those in 6–9 were as in Section 7.3.
[b] Controls include activities of ASU particles (Section 6.1.3) in the absence of F_1, OSCP, or both.
[c] Carbonylcyanide p-trifluoromethoxyphenylhydrazone, a powerful uncoupler (Section 7.3.2).
[d] 1-Anilinonaphthalene-8-sulfonate (Section 7.3.4).

by the addition of respiratory inhibitors and, curiously, stimulated by the addition of oligomycin. These reactions are lost by the addition of uncouplers irrespective of their energy sources.

7.1.5. Binding of ATPase Inhibitor to F_1

Since ATPase inhibitor is extracted by alkali, it is adsorbed on F_1 in dilute acid (Pullman and Monroy, 1963). The addition of ATP and Mg^{2+} helps the binding (Horstman and Racker, 1970).

The solution of F_1 is prepared as described in Section 7.1.3, except that a solution containing 0.25 M sucrose, 10 mM imidazole (pH 7.0), and 2 mM ATP is used to dissolve F_1 (500 μg/ml). This solution is diluted fivefold with 0.25 M sucrose and 10 mM imidazole (pH 7.0).

In a final volume of 0.2 ml, the calculated volume of 0.25 M sucrose, 20 μl F_1 (2 μg), various amounts of the inhibitor (0.05–0.3 μg), and 0.1 ml of 1 mM $MgSO_4$, 1 mM ATP, 30 mM imidazole (pH 6.7) are added in this order. After incubation at 30°C for 10 min, the ATPase activity of this mixture is assayed as described in Section 4.5. AS particles (see Section 6.1) can be used instead of pure F_1, if their specific activity is as high as 5–8 μmoles P/mg/min.

7.2. Formation of Vesicles with Intrinsic Proteins and Lipids

7.2.1. Mechanism of Vesicle Formation

Owing to the extremely low cmc (10^{-10} M) of phospholipids (Smith and Tanford, 1972), phospholipid molecules are arranged in a bilayer (Fig. 1J) when they are dispersed in a water phase (Ladbrook and Chapman, 1969). These molecules can be mobilized if some detergent is added to the lipid bilayer. *In vivo*, this mobilization is caused by a special protein (Wirtz and Zilversmit, 1968). These mobilized molecules easily interact with proteins, although some lipid-depleted hydrophobic proteins such as complexes I–IV or oligomycin-sensitive ATPase (see Section 6.3) can adsorb lipid micelles (0.5–1 mg phospholipids/mg proteins) in the absence of detergents. When the detergent is added to the phospholipids, many phases such as a cylindrical phase and an isotropic solution will appear (Small *et al.*, 1966). When these detergents are removed from phospholipids or lipid-rich lipoprotein fragments, lipid bilayers (lamellar phase) are formed again (Fig. 1J–L). A sheet of lipid bilayer is closed spontaneously to form a vesicle.

7.2.2. General Methods for Vesicle Formation

7.2.2a. Solubilization of Added Phospholipids. Phospholipid molecules must be mobilized from the highly stable bilayer by some agents such as a detergent or specific carrier protein. The amount of the detergent necessary to dissolve phospholipids in isotropic solution can be obtained from phase diagrams. For example, the phase diagram of the cholate–phosphatidyl-choline–water system (Small *et al.*, 1966) shows that at least 1 mole of sodium cholate is required to dissolve 2 moles of phosphatidylcholine, i.e., 1 g of cholate to 4 g of the lipid (see Section 2.3.3).

7.2.2b. Addition of Intrinsic Proteins. All of the intrinsic proteins required for a certain membrane function should be added with the "mobilized phospholipids" for a certain period of time to allow reorganization.

7.2.2c. Removal of the Detergents. The detergents added should be removed as completely as possible, when ion transport or phosphorylation systems are reconstituted. For this purpose, detergents with high HLB, high cmc, and low aggregation number must be used (see Section 2.3).

7.2.2d. Addition of Extrinsic Proteins. Extrinsic proteins can be added after the formation of vesicles, if these are present on the outer surface of the vesicles. For this reason, F_1 and OSCP are adsorbed on the reconstituted vesicles as described in Section 7.1.4 (Kagawa and Racker, 1971). However, extrinsic proteins should be added before the removal of detergent or the closure of the vesicles, if they are to be present on the inner surface of the vesicles. For this reason, cytochrome *c* is added during the reconstitution of third-site phosphorylation, and the remaining cytochrome *c* in the outer phase should be removed after the membrane closure (Racker and Kandrach, 1971).

7.2.3. Reconstitution of Vesicles Capable of ATP-Driven Energy Transformation

7.2.3a. Vesicles Inlayed with Oligomycin-Sensitive ATPase. These vesicles are capable of ^{32}Pi-ATP exchange, ATP-driven ion translocation, and ATP-driven formation of membrane potential (Kagawa and Racker, 1971; Kagawa, 1972*a*) (Fig. 1K,L). All of these reactions are lost by the addition of uncouplers, and ATPase activity of the vesicles is not only inhibited by oligomycin but also stimulated by the addition of uncoupler. The general method is shown in Fig. 2 and Table XI.

Table XI. Energy Transfer Activity of the Reconstituted Vesicles and Their Phospholipid Specificity[a]

Phospholipids added during the dialysis	^{32}Pi-ATP exchange activity (nmoles/mg/min)	H$^+$ accumulation driven by ATP hydrolysis (neq/mg)	ATPase activity	
			FCCP[b] stimulation (% increase)	Oligomycin inhibition (% decrease)
Phosphatidyl-choline	4.9	−2.6	24	63
Phosphatidyl-ethanolamine	17.7	−2.2	6	70
Phosphatidyl-choline plus phosphatidyl-ethanolamine	58.3	15.7	101	76

[a] The vesicles were reconstituted as described in Section 7.2.3c and pretreated with F$_1$ and OSCP as described in Section 7.3.3, ^{32}Pi-ATP exchange reaction was performed as described in Section 4.5.2, H$^+$ accumulation as described in Section 7.3.3, and ATPase activity as described in Section 4.5.1. Both phosphatidylcholine and phosphatidylethanolamine were purified from soybean phospholipid (asolectin) as described in Kagawa et al. (1972).
[b] Carbonylcyanide p-trifluoromethoxyphenylhydrazone.

7.2.3b. Solubilization of Phospholipids. The absolute requirement for cholate in this reconstitution is shown in Fig. 4. The phospholipids required in the reconstitution are the mixture of phosphatidylcholine (2 μmoles P/mg protein), phosphatidylethanolamine (2 μmoles P/mg protein), and cardiolipin (0.2 μmole P/mg protein). This mixture can be replaced by soybean phospholipid (asolectin, 5 μmoles P/mg protein) or mitochondrial phospholipid (5 μmoles P/mg protein). In all these cases, fatty acyl groups must be unsaturated (the average unsaturation between 1.2 and 2.5/acyl group). Phospholipids (20–50 nmoles P/ml) are suspended in 10 mM tricine-KOH (pH 8.0), 2% cholate, 20 mM ammonium sulfate, 25 mM sucrose, 0.1 mM EDTA, and 1 mM dithiothreitol (Kagawa et al., 1973). The suspension of the phospholipids (0.1–5 ml) is placed in a thin-walled plastic tube, filled with nitrogen, stoppered with parafilm, and sonicated for 2–5 min in a water bath at 25 °C with a Sonblaster model G201 (Narda Ultrasonic Corp., New York) at 90 W, 90 kc/sec at 1.3 A.

7.2.3c. Formation of the Vesicles. A mixture containing 1 mg of 25–40P (see Section 6.3), 4 mg sodium cholate (pH 8.0), 20 μmoles ammonium sulfate, 0.3 μmole dithiothreitol, and 5 μmoles phospholipids in a final volume of

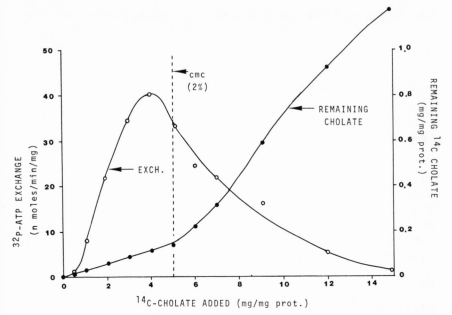

Fig. 4. Requirement for cholate during the reconstitution of vesicles capable of energy transformation. In a final volume of 0.5 ml, 2 mg of 25–40P was reconstituted at pH 8.0 with 10 mg soybean phospholipids in the presence of the indicated amount of ^{14}C-sodium cholate (1750 cpm/mg) as described in Section 7.2.3. After the dialysis, the remaining ^{14}C-cholate in the dialysate was determined by liquid scintillation counter, and ^{32}Pi-ATP exchange reaction was assayed after the addition of F_1 and OSCP as described in Section 4.5.2. From Kagawa *et al.* (1973).

0.3 ml is placed into a 1/4-inch, 0.002-inch-thick cellophane tubing. The content is dialyzed against 100 vol of solution (pH 8.0) containing 10% methanol, 1 mM dithiothreitol, 10 mM tricine-KOH, 0.1 mM ATP, 0.2 mM EDTA, and 0.2 M NaCl, at 4°C at a stirring velocity of 60 rpm for 4 hr and after a change of the dialyzing fluid for an additional 14–18 hr.

During this period, the cholate concentration is decreased ($t_{1/2}$ 150 min) as described in Section 3.5, and ^{32}Pi-ATP exchange activity of the vesicles is restored after about 10 hr dialysis (Kagawa and Racker, 1971). The increased tightness of the lipid bilayer or decrease in mobility of the fatty acyl group can also be followed by NMR spectrum (Kamat and Chapman, 1970) by the marked decrease in the signal of $(CH_2)_n$ and CH_3 (Kagawa, unpublished observation). After the dialysis, the suspension is mixed with F_1 and OSCP as described in Section 7.1 and assayed as described in Section 4.5.

7.2.4. Reconstitution of Vesicles Capable of Respiration-Driven Energy Transfer

7.2.4a. Vesicles Inlayed with Cytochrome Oxidase. The method of incorporation of cytochrome oxidase is essentially the same as the one described in the preceding section, except that the same amount of cytochrome oxidase is added instead of 25–40P. Cytochrome c is added only during the assay of the vesicles, to ensure that it is external (Hinkle *et al.*, 1972). The vesicles are capable of transporting ions in the presence of cytochrome c (see Section 7.3), and the velocity of cytochrome c oxidation is stimulated by the addition of an uncoupler.

7.2.4b. Vesicles Inlayed with Both Oligomycin-Sensitive ATPase and Cytochrome Oxidase or Complex I. The method of incorporation of these intrinsic proteins is essentially the same as the one described in the preceding section, except that these proteins as well as cytochrome c are added together. Here again, cholate is absolutely required, and phosphatidylcholine and phosphatidylethanolamine are essential. In the case of cytochrome oxidase incorporation, external cytochrome c must be removed by centrifugation of the dialysate, which is overlayered on a solution containing 0.15 M sucrose, 5 mg/ml bovine serum albumin, and 10 mM tricine-KOH (pH 7.4) at 50,000 rpm in a Spinco rotor 50Ti for 20 min. The oxidative phosphorylation of the vesicles is determined as described in Section 4.5.3 after the addition of coupling factors as described in Section 7.1.4. In the case of vesicles inlayed with complex I, special care should be taken in the preparation of the complex so as not to dialyze too much (Ragan, unpublished observation).

The preparation of the complexes and the assay of site I phosphorylation are described elsewhere (Estabrook and Pullman, 1967).

7.3. Ion Transport and Anisotropy of the Membrane

7.3.1. Vesicles and Energy Transformation

The reason why the reconstitution of oxidative phosphorylation is accompanied by the reassembly of components into vesicular membrane (Kagawa and Racker, 1971) is explained by the chemiosmotic hypothesis (Mitchell, 1966, 1967). The electrochemical potential of ions created by the electron transport system in the vesicles can be transformed into energy of ATP synthesis by the oligomycin-sensitive ATPase through which ions are carried down the potential gradient. In fact, ATP synthesis without electron transport is demonstrated if the energy is supplied in the form of a K^+ ion gradient (Cockrell *et al.*, 1967). Ion transport driven by electron transport (Mitchell, 1967; Hinkle and Horstman, 1971) or by ATP hydrolysis (*cf.* Mitchell and Moyle, 1969) has also been studied in detail.

7.3.2. Anisotropy and the Direction of Ion Movement

The membrane components are anisotropically arranged so that ions can be transported unidirectionally by the energy produced in the membrane (see Section 1.3). The direction of ion transport or the membrane potential of submitochondrial particles is opposite to that of mitochondria. This phenomenon may be caused by the topology of the components; e.g., F_1 is localized on the outer surface of submitochondrial particles, while it is on the inner surface of mitochondria (Fig. 1A–C).

In the reconstituted vesicles, if cytochrome *c* is incorporated inside or outside, the direction of ion transport by cytochrome *c* oxidation is reversed. Since ATP is impermeable to the reconstituted vesicles, the direction of ATP-driven reactions of these vesicles is the same as that in submitochondrial particles.

The reconstituted vesicles respond to uncouplers (H^+ transporter), valinomycin (K^+ transporter), nigericin (K^+–H^+ exchanger), and tetraphenyl boron (lipid-soluble anion) in the same way as submitochondrial particles and are not affected by tetraphenyl arsonium chloride (lipid-soluble cation) (Kagawa and Racker, 1971). The mechanism of action of these agents on the membranes is discussed in detail elsewhere (Haydon and Hladky, 1972; Skulachev, 1970; Montal *et al.*, 1970).

7.3.3. Proton Accumulation by ATPase in Reconstituted Vesicles (Fig. 5)

The reconstituted vesicles described in Section 7.2 are dialyzed again after the addition of F_1 and OSCP as described in Section 7.1 to remove ammonium ion from the F_1 solution. The glass electrode *p*H meter equipped with a recorder (Hinkle and Horstman, 1971) is calibrated with a standard buffer, and the full-scale range of the recorder is adjusted to 0.1 *p*H unit. Owing to the high sensitivity of the machine, electrical shielding is essential. The pretreated vesicles (0.5–1.0 mg) are put in the *p*H measuring chamber equipped with a stirrer containing 1.1 ml of 150 mM KCl, 2 mM glycylglycine buffer (*p*H 6.25), and 2 mM $MgCl_2$ at 25 °C. Then the *p*H of the suspension is adjusted to 6.25 by the addition of 0.1 N HCl from a microsyringe with stirring. ATP hydrolysis by itself releases protons if the reaction is performed at neutral *p*H, but at *p*H 6.25 in the presence of 2 mM $MgCl_2$ the reaction

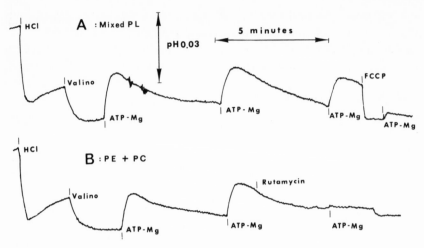

Fig. 5. Proton accumulation by ATP hydrolysis in reconstituted vesicles. A: Vesicles reconstituted with soybean phospholipids as described in Section 7.2.3; 0.85 mg. B: Vesicles reconstituted with purified phosphatidylcholine and phosphatidylethanolamine as described in Section 7.2.3. The measurement of pH was performed as described in Section 7.3.3. HCl, standard HCl, 24.2 nmoles; Valino, valinomycin, 0.2 μg; ATP-Mg, ATP–MgCl$_2$ mixture, pH 6.25, 40 nmoles each; FCCP, 2 nmoles; rutamycin, 2 μg. The medium was 150 mM KCl, 2 mM MgCl$_2$, 2 mM glycylglycine buffer, pH 6.25, 25 °C.

proceeds without net proton release (Alberty, 1968). Five minutes after the equilibration, the buffering capacity of the suspension is determined by the addition of standard HCl; 20 nequivalents gives acidification of about 0.03 pH unit. Then 0.2 μg valinomycin is added, which results in the acidification of the medium owing to H$^+$ expelled by the entrance of K$^+$ into the vesicles. After a few minutes, the addition of 40 nmoles ATP-MgSO$_4$ (pH 6.25) causes alkalinization of the medium about 0.02 pH unit/mg protein. The accumulated H$^+$ in the vesicles are released by the addition of uncoupler (Kagawa, 1972a; Kagawa *et al.*, 1973) (Fig. 5).

The electron transport–driven H$^+$ translocation of vesicles reconstituted with cytochrome oxidase has also been demonstrated (Hinkle *et al.*, 1972).

7.3.4. Fluorescence Enhancement of 1-Anilinonaphthalene-8-Sulfonate (ANS)

This method of fluorescence enhancement of ANS is convenient for detection of the "energized state" of the membrane. Like proton translocation, there are opposite effects of energization on the fluorescence of ANS with mitochondria and with submitochondrial particles. The former decrease the fluorescence and the latter enhance it. The membrane potential produced by ion translocation can expel or attract the negatively charged ANS (Jasaitis et al., 1971). The limited intramolecular rotation of the attracted ANS results in the enhanced emission of fluorescence, which is derived from excitation light energy that is usually dissipated in the intramolecular rotation.

This method can be applied to reconstituted ASU particles (Datta and Penefsky, 1970) and to vesicles (Kagawa and Racker, 1971). The fluorometer is operated with excitation light of 375 nm (Xe lamp) and emission light of 485 nm. The recorder full scale is adjusted by 0.5 μM ANS dissolved in 3 ml methanol.

The reconstituted vesicles (1 mg) pretreated with F_1 and OSCP as described in Section 7.1.4 are suspended in 3 ml of 0.05 M tris HCl (pH 7.4). To this cuvette, 6 nmoles ANS is added and then 1–5 μmoles ATP-MgSO$_4$ (pH 7.4). The increased fluorescence is destroyed by the addition of an uncoupler such as 2 μmoles FCCP.

8. REFERENCES

Ackers, G. K., 1967, A new calibration procedure for gel filtration column, *J. Biol. Chem.* **242**:3237.

Alberty, R. A., 1968, Effect of pH and metal ion concentration on the equilibrium hydrolysis of adenosine triphosphate to adenosine diphosphate, *J. Biol. Chem.* **243**:1337.

Arion, W. J., and Racker, E., 1970, Partial resolution of the enzymes catalyzing oxidative phosphorylation, XXIII. Preservation of energy coupling in submitochondrial particles lacking cytochrome oxidase, *J. Biol. Chem.* **245**:5186.

Berezney, R., Awasthi, Y. C., and Crane, F. L., 1970, The relation of phospholipid and membrane-bound ATPase in mitochondrial electron transport particles, *Bioenergetics* **1**:457.

Blair, P. V., 1967, The large scale preparation and properties of heart mitochondria from slaughterhouse material, *in* "Methods in Enzymology," Vol. 10 (R. E. Estabrook and M. E. Pullman, eds.), pp. 78–81, Academic Press, New York.

Bulos, B., and Racker, E., 1968, Partial resolution of the enzymes catalyzing oxidative phosphorylation, XVII. Further resolution of the rutamycin-sensitive adenosine triphosphatase, *J. Biol. Chem.* **243**:3891.

Burstein, C., Loyter, A., and Racker, E., 1971, Effect of phospholipase on the structure and function of mitochondria, *J. Biol. Chem.* **246**:4075.

Cattell, K. J., Lindop, C. R., Knight, L. G., and Beechey, R. B., 1971, The identification of the site of action on N,N'-dicyclohexylcarbodiimide as a proteolipid in mitochondrial membranes, *Biochem. J.* **125**:169.

Catterall, W. A., and Pedersen, P. L., 1971, Adenosine triphosphatase from rat liver mitochondria, I. Purification, homogeneity, and physical properties, *J. Biol. Chem.* **246**:4987.

Chance, B., Erecinska, M., and Lee, C. P., 1970, Localization of cytochromes in intact and fragmented mitochondrial membranes, *Proc. Natl. Acad. Sci. (U.S.)* **66**:928.

Christiansen, R. O., Loyter, A., and Racker, E., 1969, Effect of anions on oxidative phosphorylation in submitochondrial particles, *Biochim. Biophys. Acta* **18**:207.

Cockrell, R. S., Harris, E. J., and Pressman, B. C., 1967, Synthesis of ATP driven by a potassium gradient in mitochondria, *Nature* **215**:1487.

Cohn, E. J., 1925, The physical chemistry of the proteins, *Physiol. Revs.* **5**:349.

Conover, T. E., Prairie, R. L., and Racker, E., 1963, Partial resolution of the enzymes catalyzing oxidative phosphorylation, III. A new coupling factor required by submitochondrial particles extracted with phosphatides, *J. Biol. Chem.* **238**:2831.

Craig, L. C., 1960, Fractionation and characterization by dialysis, *in* "A Laboratory Manual of Analytical Methods of Protein Chemistry," Vol. 1: "Separation and Isolation of Proteins" (P. Alexander and R. J. Block, eds.), pp. 103–119, Pergamon Press, Oxford.

Datta, A., and Penefsky, H. S., 1970, Interaction of fluorescent probes with submitochondrial particles during oxidative phosphorylation, *J. Biol. Chem.* **245**:1537.

Davies, J. T., 1957, A quantitative kinetic theory of emulsion type, I. Physical chemistry of the emulsifying agent, *in* "Second International Congress on Surface Activity," Vol. I, pp. 426–438, Butterworths, London.

Davis, K. A., and Hatefi, Y., 1971, Succinate dehydrogenase, I. Purification, molecular properties, and substructure, *Biochemistry* **10**:2509.

Davis, K. A., and Hatefi, Y., 1972, Resolution and reconstitution of complex II (succinate-ubiquinone reductase) by salts, *Arch. Biochem. Biophys.* **149**:505.

Dickerson, R. E., Takano, T., Eisenberg, D., Kallai, O. B., Samson, L., Cooper, A., and Margoliash, E., 1971, Ferricytochrome *c*, I. General features of the horse and bonito proteins at 2.8 Å resolution, *J. Biol. Chem.* **246**:1511.

Estabrook, R. W., and Holowinsky, A., 1961, Studies on the content and organization of the respiratory enzymes of mitochondria, *J. Biophys. Biochem. Cytol.* **9**:19.

Estabrook, R. W., and Pullman, M. E., 1967, "Methods in Enzymology," Vol. 10: "Oxidation and Phosphorylation," pp. 1–818, Academic Press, New York.

Fisher, R. J., Chen, J. C., Sani, B. P., Kaplay, S. S., and Sanadi, D. R., 1971, A soluble mitochondrial ATP synthetase complex catalyzing ATP-phosphate and ATP-ADP exchange, *Proc. Natl. Acad. Sci. (U.S.)* **68**:2181.

Fleischer, S., and Fleischer, B., 1967, Removal and binding of polar lipid in mitochondria and other membrane systems, *in* "Methods in Enzymology," Vol. 10 (E. Estabrook and M. E. Pullman, eds.), pp. 406–433, Academic Press, New York.

Fleischer, S., Klouwen, H., and Brierley, G., 1961, Studies of the electron transfer system, XXXVIII. Lipid composition of purified enzyme preparations derived from beef heart mitochondria, *J. Biol. Chem.* **236**:2936.

Frieden, C., 1971, Protein–protein interaction and enzymatic activity, *in* "Annual Review of Biochemistry," Vol. 40, pp. 653–690, Annual Reviews, Inc., Palo Alto, Calif.

Green, D. E., 1966, The mitochondrial electron-transfer system, *in* "Comprehensive Biochemistry," Vol. 14 (M. Florkin and E. H. Stotz, eds.), pp. 309–326, Elsevier, Amsterdam.

Groot, G. S. P., Kovač, L., and Schatz, G., 1970, Promitochondria of anaerobically grown yeast, V. Energy transfer in the absence of an electron transfer chain, *Proc. Natl. Acad. Sci. (U.S.)* **68**:308.

Gulik-Krzywicki, T., Rivas, E., and Luzzati, V., 1967, Structure et polymorphisme des lipides: Etude par diffraction des rayons X du système formé de lipides de mitochondries de coeur de boeuf et d'eau, *J. Mol. Biol.* **27**:303.

Hanstein, W. G., Davis, K. A., Ghalambor, M. A., and Hatefi, Y., 1971*a*, Succinate dehydrogenase, II. Enzymatic properties, *Biochemistry* **10**:2517.

Hanstein, W. G., Davis, K. A., and Hatefi, Y., 1971*b*, Water structure and the chaotropic properties of haloacetates, *Arch. Biochem. Biophys.* **147**:534.

Haslam, J. M., Proudlock, J. W., and Linnane, A. W., 1971, Biogenesis of mitochondria, 20. The effects of altered membrane lipid composition on mitochondrial oxidative phosphorylation in *Saccharomyces cerevisiae*, *Bioenergetics* **2**:351.

Hatefi, Y., 1966, The functional complexes of the mitochondrial electron-transfer system, *in* "Comprehensive Biochemistry," Vol. 14 (M. Florkin and E. H. Stotz, eds.), pp. 199–231, Elsevier, Amsterdam.

Hatefi, Y., Haavik, A. G., Fowler, L. R., and Griffiths, D. E., 1962, Studies on the electron transfer system, XLII. Reconstitution of electron transfer system, *J. Biol. Chem.* **237**:2661.

Haydon, D. A., and Hladky, S. B., 1972, Ion transport across thin lipid membranes: A critical discussion of mechanisms in selected systems, *Quart. Rev. Biophys.* **5**:187.

Hinkle, P. C., and Horstman, L. L., 1971, Respiration driven proton transport in submitochondrial particles, *J. Biol. Chem.* **246**:6024.

Hinkle, P. C., Kim, J. J., and Racker, E., 1972, Ion transport and respiratory control in vesicles formed from cytochrome oxidase and phospholipids, *J. Biol. Chem.* **247**:1338.

Hofmann, A. F., and Small, D. M., 1967, Detergent properties of bile salts: Correlation with physiological function, *in* "Annual Review of Medicine," Vol. 18 (A. C. De Graff and W. P. Creger, eds.), pp. 333–376, Annual Reviews, Inc., Palo Alto, Calif.

Hogeboom, G. H., Schneider, W. C., and Pallade, G. E., 1948, Cytochemical studies of mammalian tissues, I. Isolation of intact mitochondria from rat liver: Some biochemical properties of mitochondria and submicroscopic particulate material, *J. Biol. Chem.* **172**:619.

Hoppel, C., and Cooper, C., 1968, The action of digitonin on rat liver mitochondria: The effect on enzyme content, *Biochem. J.* **107**:367.

Horstman, L. L., and Racker, E., 1970, Partial resolution of the enzymes catalyzing oxidative phosphorylation, XXII. Interaction between mitochondrial adenosine triphosphatase inhibitor and mitochondrial adenosine triphosphatase, *J. Biol. Chem.* **245**:1336.

Hughes, D. E., and Nyborg, W. L., 1962, Cell disruption by ultrasound, *Science* **138**:108.

Jakoby, W. B., 1971, "Methods in Enzymology," Vol. 22: "Enzyme Purification and Related Techniques," pp. 1–648, Academic Press, New York.

Jasaitis, A. A., Kuliene, V. V., and Skulachev, V. P., 1971, Anilinonaphthalenesulfonate fluorescence changes induced by non-enzymatic generation of membrane potential in mitochondria and submitochondrial particles, *Biochim. Biophys. Acta* **234**:177.

Kagawa, Y., 1967*a*, Preparation and properties of a factor conferring oligomycin sensitivity (F_0) and of oligomycin-sensitive ATPase (CF_0-F), *in* "Methods in

Enzymology," Vol. 10 (R. E. Estabrook and M. E. Pullman, eds.), pp. 505–510, Academic Press, New York.

Kagawa, Y., 1967*b*, Preparation of H-acetyl-ATPase (coupling factor 1), *in* "Methods in Enzymology," Vol. 10 (R. E. Estabrook and M. E. Pullman, eds.), pp. 526–528, Academic Press, New York.

Kagawa, Y., 1972*a*, Reconstitution of oxidative phosphorylation, *Biochim. Biophys. Acta* **265**:297.

Kagawa, Y., 1972*b*, Fractionation of membrane proteins essential for energy transfer reactions, *Fed. Proc.* **31**:416 (abst.).

Kagawa, Y., and Racker, E., 1966*a*, Partial resolution of the enzymes catalyzing oxidative phosphorylation, VIII. Properties of a factor conferring oligomycin sensitivity on mitochondrial adenosine triphosphatase, *J. Biol. Chem.* **241**:2461.

Kagawa, Y., and Racker, E., 1966*b*, Partial resolution of the enzymes catalyzing oxidative phosphorylation, IX. Reconstitution of oligomycin-sensitive adenosine triphosphatase, *J. Biol. Chem.* **241**:2467.

Kagawa, Y., and Racker, E., 1966*c*, Partial resolution of the enzymes catalyzing oxidative phosphorylation, X. Correlation of morphology and function in submitochondrial particles, *J. Biol. Chem.* **241**:2475.

Kagawa, Y., and Racker, E., 1971, Partial resolution of the enzymes catalyzing oxidative phosphorylation, XXV. Reconstitution of vesicles catalyzing ^{32}Pi-adenosine triphosphate exchange, *J. Biol. Chem.* **246**:5477.

Kagawa, Y., Takaoka, T., and Katsuta, H., 1969, Mitochondria of mouse fibroblasts, L-929, cultured in a lipid- and protein-free chemically defined medium, *J. Biochem.* **65**:799.

Kagawa, Y., Kandrach, A., and Racker, E., 1973, Partial resolution of the enzymes catalyzing oxidative phosphorylation. XXVI. Phospholipid specificity of the vesicles capable of energy transformation, *J. Biol. Chem.* **247**:676.

Kakiuchi, S., 1927, The significance of lipids in the oxygen consuming activity of tissues, *J. Biochem.* **7**:263.

Kamat, V. B., and Chapman, D., 1970, Proton magnetic resonance (PMR) spectra of erythrocyte membrane lipoproteins and apoproteins, *Chem. Phys. Lipids* **4**:323.

Keilin, D., 1930, Cytochrome and intracellular oxidase, *Proc. Roy. Soc.* **B106**:418.

Keilin, D., and Hartree, E. F., 1937, Preparation of pure cytochrome *c* from heart muscle and some of its properties, *Proc. Roy. Soc.* **B122**:298.

Knowles, A. F., and Penefsky, H. S., 1973, The subunit structure of beef heart mitochondrial adenosine triphosphatase. Physical and chemical properties of isolated subunits, *J. Biol. Chem.* **247**:6624.

Kopaczyk, K., Asai, J., Allmann, D. W., Oda, T., and Green, D. E., 1968, Resolution of the repeating unit of the inner mitochondrial membrane, *Arch. Biochem. Biophys.* **123**:602.

Ladbrook, B. D., and Chapman, D., 1969, Thermal analysis of lipids, proteins and biological membranes: A review and summary of some recent studies, *Chem. Phys. Lipids* **3**:304.

Lardy, H. A., and Ferguson, S. M., 1969, Oxidative phosphorylation in mitochondria, *in* "Annual Reviews of Biochemistry," Vol. 38, pp. 991–1034, Annual Reviews, Inc., Palo Alto, Calif.

Lenaz, G., Parenti-Castelli, G., Sechi, A. M., and Masotti, L., 1972, Lipid–protein interactions in mitochondria, III. Solvent extraction of phospholipids from mitochondrial preparations, *Arch. Biochem. Biophys.* **148**:391.

Lowry, O. H., Rosebrough, N. J., Farr, L. A., and Randall, R. J., 1951, Protein measurement with the Folin phenol reagent, *J. Biol. Chem.* **193**:265.

MacLennan, D. H., and Tzagoloff, A., 1968, Studies on the mitochondrial adenosine triphosphatase system, IV. Purification and characterization of the oligomycin sensitivity conferring protein, *Biochemistry* **7**:1603.

Mitchell, P., 1966, Chemiosmotic coupling in oxidative and photosynthetic phosphorylation, *Biol. Rev.* **41**:445.

Mitchell, P., 1967, Active transport and ion accumulation, *in* "Comprehensive Biochemistry," Vol. 22 (M. Florkin and E. H. Stotz, eds.), pp. 167–197, Elsevier, Amsterdam.

Mitchell, P., and Moyle, J., 1969, Estimation of membrane potential and pH difference across the cristae membrane of rat liver mitochondria, *Europ. J. Biochem.* **7**:471.

Montal, M., Chance, B., and Lee, C. P., 1970, Ion transport and energy conservation in submitochondrial particles, *J. Membrane Biol.* **2**:201.

Nelson, N., Deters, D. W., Nelson, H., and Racker, E., 1973, Partial resolution of the enzymes catalyzing photophosphorylation. XIII. Properties of isolated subunits of coupling factor 1 from spinach chloroplasts, *J. Biol. Chem.* **248**:2049.

Nishibayashi-Yamashita, H., Cunningham, C., and Racker, E., 1972, Resolution and reconstitution of the mitochondrial electron transport system, III. Order of reconstitution and requirement for a new factor for respiration, *J. Biol. Chem.* **247**:698.

Noguchi, T., and Freed, S., 1971, Dissociation of lipid components and reconstitution at $-75\,°C$ of Mg^{2+} dependent, Na^+ and K^+ stimulated, adenosine triphosphatase in rat brain, *Nature New Biol.* **230**:148.

Nomura, M., 1972, Assembly of bacterial ribosome, *Fed. Proc.* **31**:18.

Okunuki, K., 1960, Isolation of biologically active proteins, *in* "A Laboratory Manual of Analytical Methods of Protein Chemistry," Vol. 1 (P. A. Alexander and R. J. Block, eds.), pp. 32–64, Pergamon Press, London.

Okunuki, K., 1966, Cytochrome and cytochrome oxidase, *in* "Comprehensive Biochemistry," Vol. 14 (M. Florkin and E. H. Stotz, eds.), pp. 232–308, Elsevier, Amsterdam.

Peacocke, A. R., and Pritchard, N. J., 1968, Some biophysical aspects of ultrasound, *in* "Progress in Biophysics and Molecular Biology," Vol. 18, pp. 187–208, Pergamon Press, Oxford.

Penefsky, H. S., Pullman, M. E., Datta, A., and Racker, E., 1960, Partial resolution of the energy catalyzing oxidative phosphorylation, II. Participation of a soluble adenosine triphosphatase in oxidative phosphorylation, *J. Biol. Chem.* **235**:3330.

Porath, J., and Fornstedt, N., 1970, Group fractionation of plasma proteins on dipolar ion exchangers, *J. Chromatog.* **51**:479.

Pullman, M. E., and Monroy, G. C., 1963, A naturally occurring inhibitor of mitochondrial adenosine triphosphatase, *J. Biol. Chem.* **238**:3762.

Pullman, M. E., Penefsky, H. S., Datta, A., and Racker, E., 1960, Partial resolution of the enzymes catalyzing oxidative phosphorylation, I. Purification and properties of soluble, dinitrophenol-stimulated adenosine triphosphatase, *J. Biol. Chem.* **235**:3322.

Racker, E., 1970, "Membrane of Mitochondria and Chloroplasts," pp. 1–322, Van Nostrand Reinhold, New York.

Racker, E., and Horstman, L. L., 1967, Partial resolution of the enzymes catalyzing oxidative phosphorylation, XIII. Structure and function of submitochondrial particles completely resolved with respect to coupling factor I, *J. Biol. Chem.* **242**:2547.

Racker, E., and Kandrach, A., 1971, Reconstitution of the third site of oxidative phosphorylation, *J. Biol. Chem.* **246**:7069.

Racker, E., Fessenden-Raden, J. M., Kandrach, M. A., Lam, K. W., and Sanadi, D. R., 1970, Identity of coupling factor 2 and factor B, *Biochem. Biophys. Res. Commun.* **41**:1474.

Racker, E., Horstman, L. L., Kling, D., and Fessenden-Raden, J. M., 1969, Partial resolution of the enzymes catalyzing oxidative phosphorylation. XXI. Resolution of submitochondrial particles from bovine heart mitochondria with silicotungstate, *J. Biol. Chem.* **244**:6668.

Razin, S., 1972, Reconstitution of biological membranes, *Biochim. Biophys. Acta* **265**:241.

Reiland, J., 1971, Gel filtration, in "Methods in Enzymology," Vol. 22 (W. B. Jacoby, ed.), pp. 287-321, Academic Press, New York.

Rieske, J. S., 1967, Preparation and properties of reduced CoQ-cytochrome *c* reductase (complex IV) of the respiratory chain, in "Methods in Enzymology," Vol. 10 (R. E. Estabrook and M. E. Pullman, eds.), pp. 239–245, Academic Press, New York.

Rieske, J. S., Baum, H., Stoner, C. D., and Lipton, S. H., 1967, On the antimycin-sensitive cleavage of complex III of the mitochondrial respiratory chain, *J. Biol. Chem.* **242**:4854.

Ryrie, I. J., and Jagendorf, A. T., 1972, Correlation between a conformational change in the coupling protein and high energy state in chloroplasts, *J. Biol. Chem.* **247**:4453.

Sachs, D. H., and Painter, E., 1971, Improved flow rates with porous Sephadex gel, *Science* **175**:782.

Schneider, D. L., Kagawa, Y., and Racker, E., 1972, Chemical modification of the inner mitochondrial membrane, *J. Biol. Chem.* **247**:4074.

Senior, A. E., 1971, On the relationship between the oligomycin-sensitivity conferring protein and other mitochondrial coupling factors, *Bioenergetics* **2**:141.

Singer, T. P., 1966, Flavoprotein dehydrogenases of the respiratory chain, in "Comprehensive Biochemistry," Vol. 14 (M. Florkin and E. H. Stotz, eds.), pp. 127–198, Elsevier, Amsterdam.

Skulachev, V. P., 1970, Electric fields in coupling membranes, *FEBS Letters* **11**:301.

Slater, E. C., 1966, Oxidative phosphorylation, in "Comprehensive Biochemistry," Vol. 14 (M. Florkin and E. H. Stotz, eds.), pp. 327–396, Elsevier, Amsterdam.

Slater, E. C., 1971, The coupling between energy-yielding and energy-utilizing reactions in mitochondria, *Quart. Rev. Biophys.* **4(I)**:35.

Small, D. M., Bourgès, M. C., and Dervichian, D. G., 1966, The biophysics of lipidic associations, I. The ternary systems lecithin–bile salt–water, *Biochim. Biophys. Acta* **125**:563.

Smith, R., and Tanford, C., 1972, The critical micelle concentration of L-α-dipalmitoylphosphatidylcholine in water and water/methanol solutions, *J. Mol. Biol.* **67**:75.

Sober, H. A., 1970, "Handbook of Biochemistry," 2nd ed., pp. A1-In.108, Chemical Rubber Co., Cleveland, Ohio.

Stekhoven, F. S., 1972, Energy transfer factor A.D (ATP synthetase) as a complex Pi-ATP exchange enzyme and its stimulation by phospholipids, *Biochem. Biophys. Res. Commun.* **47**:7.

Stekhoven, F. S., Waitkus, R. F., and van Moerkerk, H. T. B., 1972, Identification of the dicyclohexylcarbodiimide-binding protein in the oligomycin-sensitive adenosine triphosphatase from bovine heart mitochondria, *Biochemistry* **11**:1144.

Swanljung, P., 1971, Detergent gradient gel chromatography: A technique for purification of membrane-bound enzymes, *Anal. Biochem.* **43**:382.

Tanford, C., 1972, Hydrophobic free energy, micelle formation and the association of proteins with amphiphiles, *J. Mol. Biol.* **67**:59.

Tsuneyoshi, K., 1927, Effect of lipin and allied substances on the oxidative activity of tissues, *J. Biochem.* **7**:235.

Tzagoloff, A., MacLennan, D. H., McConnell, D. G., and Green, D. E., 1967, Studies on the electron transfer system, LXVIII. Formation of membranes as the basis of the reconstitution of the mitochondrial electron transfer system, *J. Biol. Chem.* **242**:2051.

Tzagoloff, A., Byington, K. H., and MacLennan, D. H., 1968*a*, Studies on the mitochondrial adenosine triphosphatase system, II. The isolation and characterization of an oligomycin-sensitive adenosine triphosphatase from bovine heart mitochondria, *J. Biol. Chem.* **243**:2405.

Tzagoloff, A., McConnell, D. G., and MacLennan, D. H., 1968*b*, Studies on the electron transfer system, LXIX. Solubilization of the mitochondrial inner membrane by sonic oscillation, *J. Biol. Chem.* **243**:4117.

Uesugi, S., Dulak, N. C., Dixon, J. F., Hexum, T. D., Dahl, J. L., Perdue, J. F., and Hokin, L. E., 1971, Studies on the characterization of the sodium–potassium transport adenosine triphosphatase, VI. Large scale partial purification and properties of a lubrol-solubilized bovine brain enzyme, *J. Biol. Chem.* **246**:531.

van Dam, K., and Meyer, A. J., 1971, Oxidation and energy conservation by mitochondria, *in* "Annual Review of Biochemistry," Vol. 40, pp. 115–160, Annual Reviews, Inc., Palo Alto, Calif.

van de Stadt, R. J., Kraaipoel, R. J., and van Dam, K., 1972, $F_1 \cdot X$. A complex between F_1 and OSCP, *Biochim. Biophys. Acta* **267**:25.

Wainio, W. W., 1970, "The Mammalian Mitochondrial Respiratory Chain," pp. 1–499, Academic Press, New York.

Weber, K., and Kuter, D. J., 1971, Reversible denaturation of enzymes by sodium dodecyl sulfate, *J. Biol. Chem.* **246**:4504.

Weber, K., and Osborn, M., 1969, The reliability of molecular weight determination by dodecyl sulfate–polyacrylamide gel electrophoresis, *J. Biol. Chem.* **244**:4406.

Weiss, H., and Bücher, T., 1970, Chromatographic separation of membrane proteins on lipophilic ion exchange resins: The influence of various resin-linked aliphatic chains, *Europ. J. Biochem.* **17**:561.

Wirtz, K. W. A., and Zilversmit, D. B., 1968, Exchange of phospholipids between liver mitochondria and microsome *in vitro*, *J. Biol. Chem.* **243**:3596.

Yakushiji, E., and Okunuki, K., 1940, Ueber eine neue Cytochrom-komponente und ihre Funktion, *Proc. Imp. Acad.* **16**:229.

Zahler, P., and Weibel, E. R., 1970, Reconstitution of membranes by recombining proteins and lipids derived from erythrocyte stroma, *Biochim. Biophys. Acta* **219**:320.

Index

271